ANTENTOP

ANTENTOP 01 2014 # 018

ANTENTOP is *FREE* e-magazine devoted to ANTENna's

1-2014

Theory,

Operation, and

Practice

Edited by hams for hams

In the Issue:
Antennas Theory!

Practical design of HF Antennas!

Underground Antennas!

Practical design of VHF Antennas!

Regenerative Receivers!

Lawn Antenna *K2MIJ*

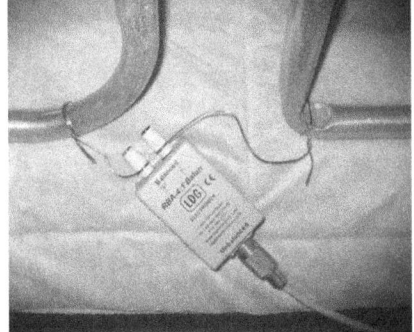

Thanks to our authors:

Prof. Natalia K.Nikolova

Nick Kudryavchenko, UR0GT

Aleksandr Grachev, UA6AGW

Vladimir Fursenko, UA6CA

And others…..

Pocket Antenna Tuner by *UN7CI*

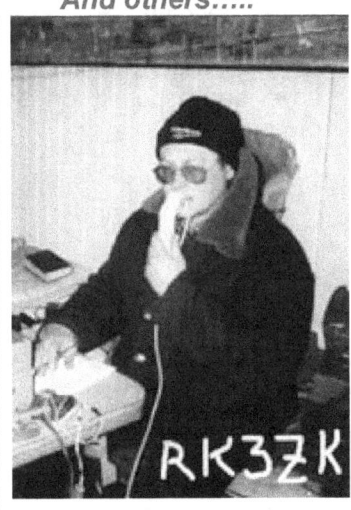

EDITORIAL:
Well, my friends, new ANTENTOP – 01 - 2014 come in! ANTENTOP is just authors' opinions in the world of amateur radio. I do not correct and re-edit yours articles, the articles are printed "as are". A little note, I am not a native English, so, of course, there are some sentence and grammatical mistakes there… Please, be indulgent! ANTENTOP 01 –2014 contains antenna articles, description of antenna patents, Regenerative Receivers. Hope it will be interesting for you.
Our pages are opened for all amateurs, so, you are welcome always, both as a reader as a writer.

73! Igor Grigorov, VA3ZNW

ex: RK3ZK, UA3-117-386,

UA3ZNW, UA3ZNW/UA1N, UZ3ZK

op: UK3ZAM, UK5LAP,

EN1NWB, EN5QRP, EN100GM

Copyright: Here at ANTENTOP we just wanted to follow traditions of FREE flow of information in our great radio hobby around the world. *A whole issue* of ANTENTOP may be photocopied, printed, pasted onto websites. We don't want to control this process. It comes from all of us, and thus it belongs to all of us. This doesn't mean that there are no copyrights.

There is! *Any work is copyrighted by the author. All rights to a particular work are reserved by the author.*

Contact us: Just email me or drop a letter.
Mailing address:
59- 547 Steeles Ave West., Toronto, ON, M2M3Y1, CANADA
Or mail to:antentop@antentop.org
NB: Please, use only plain text and mark email subject as: **igor_ant**. I receive lots spam, so, I delete **ALL** unknown me messages **without** reading.

ANTENTOP is *FREE* e-magazine, available *FREE* at http://www.antentop.org/

Editorial

Welcome to ANTENTOP, FREE e - magazine!

ANTENTOP is *FREE e- magazine*, made in *PDF*, devoted to Antennas and Amateur Radio. Everyone may share his experience with others hams on the pages. Your opinions and articles are published without any changes, as I know, every your word has the mean.

Every issue of ANTENTOP is going to have 100 pages and this one will be paste in whole on the site. I do not know what a term for one issue would be taken, may be 12 month or so. A whole issue of ANTENTOP holds nearly 10- 30 MB.

A little note, I am not native English, so, of course, there are some sentence and grammatical mistakes there… Please, be indulgent!

Publishing: If you have something for share with your friends, and if you want to do it *FREE*, just send me an email. Also, if you want to offer for publishing any stuff from your website, you are welcome!

Your opinion is important for me, so, contact if you want to say something!

Copyright Note:

Dear friends, please, note, I respect Copyright. Always, when I want to use some stuff for ANTENTOP, I ask owners about it. But… sometimes my efforts are failed. I have some very interesting stuff from closed websites, but I failed go to touch with their owners… as well as I have no response on some my emails from some owners.

I have a big collection of pictures. I have got the pictures and stuff in different ways, from *FREE websites*, from commercial CDs, intended for *FREE using*, and so on… I use to the pictures (and seldom, some stuff from free and closed websites) in ANTENTOP. *If the owners of the Copyright stuff are have concern*, please, contact with me, I immediately remove any Copyright stuff, or, if it is necessary, all needed references will be made there.

Business Advertising: *ANTENTOP is not a commercial magazine.* Authors and I (Igor Grigorov, the editor of the magazine) do not get any profit from any issue. But of course, I do not mention from commercial ads in ANTENTOP. It allows me to do the magazine in most great way, allows me to pay some money for authors to compensate their hard work.

So, if you want paste a commercial advertisement in ANTENTOP, please contact me. A commercial advertisement will do ANTENTOP even greater interesting and various!

Book Advertising: I believe that *Book Advertising* is a noncommercial advertisement. So, Book Advertising is *FREE* at ANTENTOP. Contact with me for details.

Email: igor.grigorov@gmail.com
subject: igor_ant

NB: Please, use only plain text and mark email subject as: *igor_ant*. I receive lots spam and viruses, so, I delete *ALL unknown me messages* without reading.

73! *Igor Grigorov*, VA3ZNW
ex: UA3-117-386, UA3ZNW, UA3ZNW/UA1N, UZ3ZK, RK3ZK
op: UK3ZAM, UK5LAP, EN1NWB, EN5QRP, EN100GM

http://www.antentop.org/

Table of Contents

Antenna Theory

#	Title	Page
1	**Linear Array Theory- Part II : by: Prof. Natalia K. Nikolov** Dear friends, I would like to give to you an interesting and reliable antenna theory. Hours searching in the web gave me lots theoretical information about antennas. Really, at first I did not know what information chose for ANTENTOP. Now I want to present to you one more very interesting Lecture - it is LECTURE 16: Linear Array Theory- Part II Hansen-Woodyard end-fire array, directivity of a linear array, linear array pattern characteristics – recapitulation; 3-D characteristics of an N-element linear array...	5- 19

HF- Antenna Practice

#	Title	Page
2	**44 Feet Dipole. Where is the Truth?: by: Arnie Coro, CO2KK, CUBA** After more than half a century working with professional and amateur radio antenna systems... All I can say this is a never ending subject!! The 44 feet long dipole, 22 feet on each leg came out of the work of the prematurely deceased Len Cebik W4RNL...	20- 24
3	**Lawn Antenna: by: James R Kellner, K2MIJ** Ok Gang in case you don't remember me....I am the guy who ran a couple of "Forks" as a dipole awhile back and actually worked W1AW/5 Oklahoma running 5 watts SSB from my trusty Yaesu FT-817....Well I have a new one for you and help will be appreciated to reach my goal!... .	25- 26
4	**Buried Antennas for Emergency Communications: by: John J. Schultz W1DCG/W2EEY: Credit Line: 73 MAGAZINE, April 1967, pp.: 34- 35** In this article, W1DCG describes some of the properties of buried antennas, particularly in relation to their usefulness for amateur Civil Defense or emergency communications installations....	27- 29
5	**Simple Broadband Antenna for the 40- meter Band: by: Igor Grigorov, va3znw** The Simple Broadband Antenna was designed on base of my local environment and taking into account the ease/cheap to do...	30- 34
6	**Directional Antenna UA6AGW V. 7.00: by: Aleksandr Grachev, UA6AGW** The antenna was born after numerous experiments that were made in past three years. Russian Patent # 125777 was obtained for the antenna...	35- 39

Table of Contents

		Page
7	**Antenna UA6AGW in Experimenters by RU1OZ : by: Nikolay Chabanov, RU1OZ**	40- 41
	In my native city Archangelsk my house is located very close to the Power Transmission Line. High noise from the line forced me to use Magnetic Loop Antennas. The antennas could work very effective and may eliminate the electrical noise. I can make QSOs with EU and JA (at 10- 18- MHz) using only 4-watts with the Magnetic Loop Antenna. But... Magnetic Loop Antenna has some disadvantages for me...	
8	**Antenna UA6AGW V.40.20: by: Aleksandr Grachev, UA6AGW**	42
	The experimental antenna made on the base of previously described versions of the Antenna UA6AGW for 40- meter Band. Aim of the experiment was to reduce the room that the horizontal wire takes ...	
9	**Field Antenna UA6AGW V.40.21 : by: Aleksandr Grachev, UA6AGW**	43- 46
	The antenna was designed for installation in a field conditions or limited space. Antenna may be installed at a low- height mast. Antenna does not required guys and takes small room for installation...	
10	**Shortened Antenna G5RV for 14- 50- MHz Bands: by: Alex Karakakan, UY5ON, Kharkov Ukraine**	47
	Shortened antenna G5RV is a variant classical G5RV with shortened radiation parts and shortened matching two-wire line. Amateur's Bands within 14- 50- MHz spread are mostly welcome for DX operation. So in most cases the antenna could provide good operation with DX- stations. Antenna takes small room...	
11	**Shortened Dipole Balcony Antenna for the 20- meter Band: by: Viktor Kovalensky, RN9AAA**	48- 50
	Some years ago, just for fun, I made the Shortened Dipole Antenna for the 20- meter Band at my balcony. The antenna still exists and I use to it for my operation in the Air...	

Table of Contents

		Page
	Simple Window Loop Antenna: by: Aleksandr Sterlikov, RA9SUS	
12	The antenna was installed across wooden window frame. Perimeter of the loop in my case was 4.7- meter. Antenna could be tuned from 14 to 30- MHz. I fed the antenna 50- Ohm coaxial cable. However, also I used to 75- Ohm coaxial cable with success......	51
	Simple Folded Dipole Antenna for the 20- meter Band: by: Vladimir E. Tokarev, UA4HAZ	
13	The antenna was designed for limited space. It may be installed on balcony, on fence or at backyard and masked on to rope for drying clothes. Antenna has input impedance cloth to 50- Ohm...	52- 53
	Simple Wire Antenna for All HF- Bands : by: Vladimir Fursenko, UA6CA	
14	The simple wire antenna works well from 160- to 10 meter. The antenna may be tuned (to needed amateur band) at a shack. Antenna contains only one tuning parts- it is a variable capacitor 10- 200- pF. An inductor (near 3... 5- micro- Henry) is switched in serial with the antenna at the 40- meter Band...	54
	Twin Triangle Antenna for the 10- meter Band: by: Yuriy Kondrat'ev, UA1ZAS	
15	The antenna is used at the 10- meter Band. Antenna made from two wires triangles. The triangles are fastened to my Ground Plane antenna for the 20- meter band. The triangles do not influenced to the ground plane ...	55
	Compact Twin Delta Antenna for the 80- and 40- meter Bands: by: John J. Schultz, W2EEY/7	
16	The antenna has wide broadband at the 80- and 40- meter Bands. So the antenna does not require any tuning. Antenna is simple in design and takes lots room for installation. Antenna radiates a vertical polarisation wave with almost circle DD in horizontal plane. Antenna is not critical to the sizes. It could be got good result at perimeter of each triangle lambda/4 at lower band at the antenna....	55
	Delta Antenna for 80-. 40-, 20- and 15- meter Bands: By: Nikolay Kudryavchenko, UR0GT	
17	The Delta Antenna has perimeter 86- meter. Antenna has resonances at four amateur bands. There are 80, 40, 20 and 15 meter. However, input impedance at the bands not allows use a 50- Ohm coaxial cable to feed the antenna with low SWR at all the bands. Best solution is to use a 100- Ohm coaxial cable....	56- 58

Table of Contents

Page		Page
18	**Windom UA6CA for 80-, 40-, 20- and 10- meter Bands: by: Vladimir Fursenko, UA6CA** Antenna was installed at the edge of the roof of the 5- store house. Transmitter was placed at the first floor. For improving of the efficiency of the antenna a grounding "mirror" wire was dug in the ground. Mirror wire was in plastic insulation. Ends of the mirror wire and connection to the wire were insulated from the ground. ...	59
19	**Air Plane HF Antennas** Some pictures of real HF-Antennas installed at Air Planes and Helicopters...	60

VHF- UHF Antennas

20	**Car Antenna for 435- MHz : by: Nikolay Kudryavchenko, UR0GT** The Car Antenna for the 435- MHz has very good parameters- circle DD in Horizon Plane and small angle lobe at Vertical Plane. Antenna has input impedance 50- Ohm. Antenna has wide pass band.....	61- 63
21	**Antenna for 2-meter Band, LPD (433), 70- cm Band and for RMR (446): by: Igor Vakhreev, RW4HFN** It is very simple antenna that allows works at several frequencies bands with low SWR. The antenna is enough broadband that does not required hold strictly sizes at the design. ...	64-66
22	**Twin Delta Antenna for the 2- meter Band: by: Nikolay Kudryavchenko, UR0GT** The simple Twin Delta Antenna works fine at the 2- meter Band. The antenna is enough broadband that does not required hold strictly sizes at the design. Antenna could be made from wide range diameters of wire - 2... 10- mm would be good. Antenna does not require any symmetrical devices....	67-68
23	**Conversion Auto CB-Antenna HUSTLER-1C-100 to Antenna for the 2- meter Band : by: Igor Mishin, UT3IM** It is very easy convert Auto CB-Antenna HUSTLER - 1C-100 (on magnet base) to antenna working at 2- meter Band. Figure 1 shows the conversation.....	69

Table of Contents

Page		Page
24	**UB5UG Snake Antenna : by: Yuri Medinets, UB5UG** Note from I.G.: The Snake Antenna was very popular in the ex-USSR. The antenna came to ham from a page of paper with hand writer schematic. The schematic was introduced by Yuri, UB5UG. At first, the antenna was widely used at Ukraine then it came to other republics of the ex-USSR. It was a very simple antenna that very easy could be made from a coaxial cable. The antenna could be very easy redesign for other (as well for TV) bands.	70
25	**Horizontal Antenna with Vertical Polarization for the 2- meter Band : by: Vasiliy Oleynik, RW4HX** At my vacation I guested my friends at their cottage near 40-km from the city. On the second day I decided to try my FT51. Oops, nobody can copy me when I transmitted on to the transceiver's rubber duck. So I need an antenna that could take the 40 km.	71- 73

TV Antennas

26	**Broadband TV Antenna: By: Nikolay Kudryavchenko, UR0GT** The simple broadband TV antenna works at the 580- 760- MHz. Passband of the antenna is 180- MHz. Antenna has input impedance 300- Ohm at the pass band. Antenna may be used with antenna amplifier that has such input impedance. Antenna may be used with coaxial cable with broadband transformer.	74- 75
27	**Chireix- Mesny TV Antenna: By: Nikolay Kudryavchenko, UR0GT** The simple broadband TV antenna works at the 615- 765- MHz. Antenna has input impedance 300- Ohm at the pass band. Antenna may be used with antenna amplifier that has such input impedance.	76- 77

Useful Pieces

28	**Tube Socket from Surplus VHF Resonator: by: Robert Akopov, UN7RX** Simple Home Brew Power Tube Socket from Surplus VHF Resonators....	78

Table of Contents

Page

Tube Transceiver

Regenerative Transceiver for 160- meter Band: by: Georgiy Gorelashvili, 4L1G

29

There is experimental AM transceiver that may be used at local communication. Practically any pentode would work at the circuit. Frequency of the transmitter is not stable because antenna is switched on directly to the oscillator's inductor. Regime of the regenerative receiver could not be optimal because of high coupling of the antenna with receiver's inductor. However, at small antenna length - in 2... 5 meter and small distances the transceiver is quite well for experimental work...

79

Tube Transmitters

Retro AM Tube Transmitter for 160-meter: by: Georgiy Gorelashvili, 4L1G

30

The schematic (in different variations) was very popular in ex-USSR. There is possible use any powerful pentode with tap from the third grid (for modulation). Of course, frequency of the transmitter is not stable but it is quite possible use for experimental purposes....

80

CW Tube Transmitter for 40 and 80- meter Band: by: Georgiy Gorelashvili, 4L1G

31

The schematic (in different variations) was very popular in ex-USSR. There is possible use any powerful tetrode for the rig. (Note from I.G.: I also used such rig in 70s- 80s years. The rig works fine. I used pentode 6P15P, 6P14P, 6P3S at the transmitter.)....

81

Simple AM Tube Transmitter for 1.5- 3.5- MHz: by: Georgiy Gorelashvili, 4L1G

32

The schematic (in different variations) was very popular in ex-USSR. There is possible use any powerful pentode with tap from the third grid (for modulation). Of course, frequency of the transmitter is not stable but it is quite possible use for experimental purposes........

82

Regenerative Receivers

Simple MW- HF- Regenerative Receiver: by: Seiji: Credit Line: Forum at www.cqham.ru

33

Simple MW- HF Regenerative receiver has two independently stages- one for MW-Band another one for HF-Band....

83

Table of Contents

		Page
34	**Simple Regenerator Receiver with Loop Antenna:** by: Aleksandr Bulanenko, UA6AAK It is next generation of simple battery powered regenerative receiver. The receiver used to the input inductor as antenna. It was made two inductors, one for 3.3- 14.6- MHz another one for 4.0- 19.0- MHz...	84

Antenna Tuning Units

35	**Simple HF ATU on Lengths of Coaxial Cable:** by: Igor Grigorov, va3znw The ATU was made by me in far 80s. It was may be a simplest ATU what I made ever. It contains only one rotary switch and rolls of a coaxial cable. But the ATU works very well. The ATU has only one lack- sizes. Sizes of the ATU are not small. Below there are several words to the theoretical base of the ATU...	85- 88
36	**Converting Antenna Tuner MFJ-962D for Operation with Symmetrical Ladder Line:** By: Viktor Drobot, RK3DL For operation in the Air at all HF- Bands I use to antenna Delta. The antenna is fed by 300- Ohm Ladder Line. To match the antenna with my transceiver I use to ATU MFJ-962D. The ATU has symmetrical transformer at output. The transformer could provide good symmetrical operation ... but with antennas that has low reactance. My Delta has significant reactance through amateur's bands. So the concept is not for me....	89
37	**Pocket Antenna Tuner** This article is described a small (almost pocket) Antenna Tuner that can work with 100- Watt transceiver....	90- 92

Books

38	**GROUNDING, BONDING, AND SHIELDING** Excellent Reference to Grounding, Bounding and Shielding. The two books are covered lots practical questions. It is really good books that you can find in the Internet.....	93

Page 4.3   http://www.antentop.org/

Propagation

Bridge Effect: by: Igor Grigorov, va3znw

39 — Every day when I drive my car to and from my job I drive under bridges and arches. — 94- 96

Inside my car I usually listen to radio. My favorite radio is 680 News Radio. The radio station works on 680- kHz. This radio transmits useful news for me. It's the weather, what is and what will be news in Toronto and the World, as well as local traffic, which road is open, what the road is closed due to an accident construction. Knowledge of the traffic saves me a lot of time. I noticed that volume of the 680-News usually is changed when I drove under a bridge or under an arch.....

Noise and Hum

Nature of Hum : by: Michael J Hebert, NH7SR

40 — Nature of Hum that harm to DC or regenerative receiver. What is the question that is stay before a ham who try such type of receivers.... — 97

PATENTS

Antenna for Mobile Communications: by: A.G. Kandoian

41 — Description of Patent of a Patent for Antenna for Mobile Communications. — 98- 100

Обеспечение питанием радиостанции в полевых условиях.

Antenna Theory

www.antentop.org

Feel Yourself a Student!

Dear friends, I would like to give to you an interesting and reliable antenna theory. Hours searching in the web gave me lots theoretical information about antennas. Really, at first I did not know what information to choose for ANTENTOP. Finally, I stopped on lectures "Modern Antennas in Wireless Telecommunications" written by Prof. Natalia K. Nikolova from McMaster University, Hamilton, Canada.

You ask me: Why?

Well, I have read many textbooks on Antennas, both, as in Russian as in English. So, I have the possibility to compare different textbook, and I think, that the lectures give knowledge in antenna field in great way. Here first lecture "Introduction into Antenna Study" is here. Next issues of ANTENTOP will contain some other lectures.

So, feel yourself a student! Go to Antenna Studies!

I.G.

My Friends, the above placed Intro was given at ANTENTOP- 01- 2003 to Antennas Lectures.

Now I know, that the Lecture is one of popular topics of ANTENTOP. Every Antenna Lecture was downloaded more than 1000 times!

Now I want to present to you one more very interesting Lecture 16- it is a Lecture Linear Array Theory- Part II. I believe, you cannot find such info anywhere for free! Very interesting and very useful info for every ham, for every radio- engineer.

So, feel yourself a student! Go to Antenna Studies!

I.G.

McMaster University Hall

Prof. Natalia K. Nikolova

Linear Array Theory- Part II

Linear arrays: Hansen-Woodyard end-fire array, directivity of a linear array, linear array pattern characteristics – recapitulation; 3-D characteristics of an N-element linear array. ...
by Prof. Natalia K. Nikolova

LECTURE 16: LINEAR ARRAY THEORY - PART II

(Linear arrays: Hansen-Woodyard end-fire array, directivity of a linear array, linear array pattern characteristics – recapitulation; 3-D characteristics of an N-element linear array.)

1. Hansen-Woodyard end-fire array (HWEFA)

   One of the shortcomings of end-fire arrays (EFA) is their relatively broad HPBW as compared to broadside arrays.

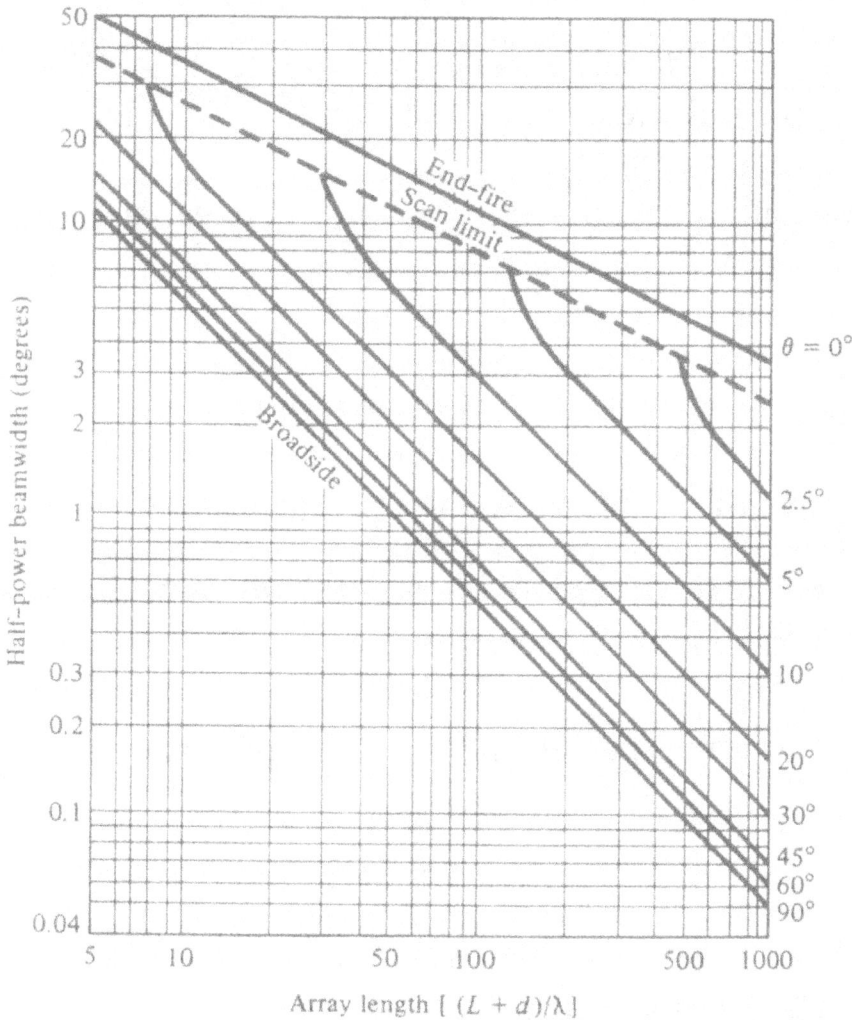

Half-power beamwidth for broadside, ordinary end-fire, and scanning uniform linear arrays. (SOURCE: R. S. Elliott, "Beamwidth and Directivity of Large Scanning Arrays," First of Two Parts, *The Microwave Journal*, December 1963)

Fig. 6-11, pp.270, Balanis

To enhance the directivity of an end-fire array, Hansen and Woodyard proposed that the phase shift of an ordinary EFA
$$\beta = \pm kd$$
be increased as:

$$\beta = -\left(kd + \frac{2.94}{N}\right) \text{ for maximum at } \theta = 0° \quad (16.1)$$

$$\beta = +\left(kd + \frac{2.94}{N}\right) \text{ for maximum at } \theta = 180° \quad (16.2)$$

Conditions (16.1)–(16.2) are known as the Hansen – Woodyard conditions for end-fire radiation. They follow from a procedure for maximizing the directivity.

The normalized pattern AF_n of a uniform linear array is:

$$AF_n \approx \frac{\sin\left[\frac{N}{2}(kd\cos\theta + \beta)\right]}{\frac{N}{2}(kd\cos\theta + \beta)} \quad (16.3)$$

if the argument $\psi = \frac{N}{2}(kd\cos\theta + \beta)$ is sufficiently small (see previous lecture). We are looking for optimal β, which would result in maximum directivity. Let:

$$\beta = -pd \quad (16.4)$$

where d is the array spacing and p is the optimization parameter

$$\Rightarrow AF_n = \frac{\sin\left[\frac{Nd}{2}(k\cos\theta - p)\right]}{\frac{Nd}{2}(k\cos\theta - p)}$$

Assume that $Nd/2 = q$; then

$$\Rightarrow AF_n = \frac{\sin\left[q(kd\cos\theta - p)\right]}{q(kd\cos\theta - p)} \quad (16.5)$$

or $AF_n = \dfrac{\sin Z}{Z}$; where $Z = q(kd\cos\theta - p)$

The radiation intensity:

$$U(\theta) = |AF_n|^2 = \dfrac{\sin^2 Z}{Z^2} \quad (16.6)$$

$$U(\theta = 0) = \left\{\dfrac{\sin[q(k-p)]}{q(k-p)}\right\}^2 \quad (16.7)$$

$$U_n(\theta) = \dfrac{U(\theta)}{U(\theta=0)} = \left\{\dfrac{z}{\sin z}\dfrac{\sin Z}{Z}\right\}^2 \quad (16.8)$$

where:

$z = q(k - p)$
$Z = q(k\cos\theta - p)$, and
$U_n(\theta)$ - normalized power pattern with respect to $\theta = 0°$.

Directivity at $\theta = 0°$:

$$D_0 = \dfrac{4\pi U(\theta = 0)}{P_{rad}} \quad (16.9)$$

where $P_{rad} = \oiint_\Omega U_n(\theta) d\Omega$. To maximize the directivity, the quantity $U_0 = P_{rad}/4\pi$ will be minimized.

$$U_0 = \dfrac{1}{4\pi}\int_0^{2\pi}\int_0^{\pi}\left(\dfrac{z}{\sin z}\dfrac{\sin Z}{Z}\right)^2 \sin\theta d\theta d\theta \quad (16.10)$$

$$U_0 = \dfrac{1}{2}\left(\dfrac{z}{\sin z}\right)^2 \int_0^{\pi}\left\{\dfrac{\sin[q(k\cos\theta - p)]}{q(k\cos\theta - p)}\right\}^2 \sin\theta d\theta \quad (16.11)$$

$$U_0 = \dfrac{1}{2kq}\left(\dfrac{z}{\sin z}\right)^2 \left[\dfrac{\pi}{2} + \dfrac{\cos 2z - 1}{2z} + \text{Si}(2z)\right] = \dfrac{1}{2kq}g(z) \quad (16.12)$$

Here, $\text{Si}(z) = \int_0^z \frac{\sin t}{t} dt$. The minimum of $g(z)$ occurs when

$$z = q(k-p) = -1.47 \quad (16.13)$$

$$\Rightarrow \frac{Nd}{2}(k-p) = -1.47$$

$$\Rightarrow \frac{Ndk}{2} - \frac{Ndp}{2} = -1.47, \quad \text{where} \quad dp = -\beta$$

$$\Rightarrow \frac{N}{2}(dk + \beta) = -1.47$$

$$\boxed{\beta = -\frac{2.94}{N} - kd = -\left(kd + \frac{2.94}{N}\right)} \quad (16.14)$$

Equation (16.14) gives Hansen-Woodyard condition for improved directivity along $\theta = 0°$. Similarly, for $\theta = 180°$:

$$\boxed{\beta = +\left(\frac{2.94}{N} + kd\right)} \quad (16.15)$$

Usually, conditions (16.14) and (16.15) are approximated by:

$$\beta = \pm\left(kd + \frac{\pi}{N}\right) \quad (16.16)$$

which is easier to remember and gives almost identical results since the curve $g(z)$ at its minimum is very flat.

Conditions (16.14)-(16.15), or (16.16), ensure minimum beamwidth (maximum directivity) in the desired end-fire direction but there is a trade-off in the side-lobe level, which is higher than that of the ordinary EFA. Besides, conditions (16.14)-(16.15) have to be complemented by additional requirements, which would ensure low level of the radiation in the direction opposite to the main lobe.

a) For a maximum at $\theta = 0°$:

$$\beta = -\left(kd + \frac{2.94}{N}\right)\Bigg|_{\theta=0°} \Rightarrow \begin{array}{l} \psi_{\theta=0°} = -\dfrac{2.94}{N} \\ \psi_{\theta=180°} = -2kd - \dfrac{2.94}{N} \end{array} \quad (16.17)$$

In order to have a minimum of the pattern in the $\theta = 180°$ direction, one must ensure that:

$$|\psi|_{\theta=180°} \simeq \pi, \quad (16.18)$$

It is easier to remember Hansen-Woodyard conditions for maximum directivity in the $\theta = 0°$ direction as:

$$\begin{array}{l} |\psi|_{\theta=0°} = \dfrac{2.94}{N} \simeq \pi \\ |\psi|_{\theta=180°} \simeq \pi, \end{array} \quad (16.19)$$

b) For a maximum at $\theta = 180°$:

$$\beta = kd + \frac{2.94}{N}\Bigg|_{\theta=180°} \Rightarrow \begin{array}{l} \psi_{\theta=180°} = \dfrac{2.94}{N} \\ \psi_{\theta=0°} = 2kd + \dfrac{2.94}{N} \end{array} \quad (16.20)$$

In order to have a minimum of the pattern in the $\theta = 0°$ direction, one must ensure that

$$|\psi|_{\theta=0°} \simeq \pi, \quad (16.21)$$

One can now summarize Hansen-Woodyard conditions for maximum directivity in the $\theta = 180°$ direction as:

$$\begin{array}{l} |\psi|_{\theta=180°} = \dfrac{2.94}{N} \simeq \dfrac{\pi}{N} \\ |\psi|_{\theta=0°} \simeq \pi \end{array} \quad (16.22)$$

If (16.18) and (16.21) are not observed, the radiation in the opposite of the desired direction might even exceed the main beam level. It is easy to show that the complimentary requirement of

$|\psi| = \pi$ at the opposite direction can be met, if the following relation is observed:

$$d = \left(\frac{N-1}{N}\right)\frac{\lambda}{4} \qquad (16.23)$$

If N is large, $d \simeq \lambda/4$. Thus, for a large uniform array, Hansen-Woodyard condition can yield improved directivity, only if the spacing between the array elements is approximately $\lambda/4$.

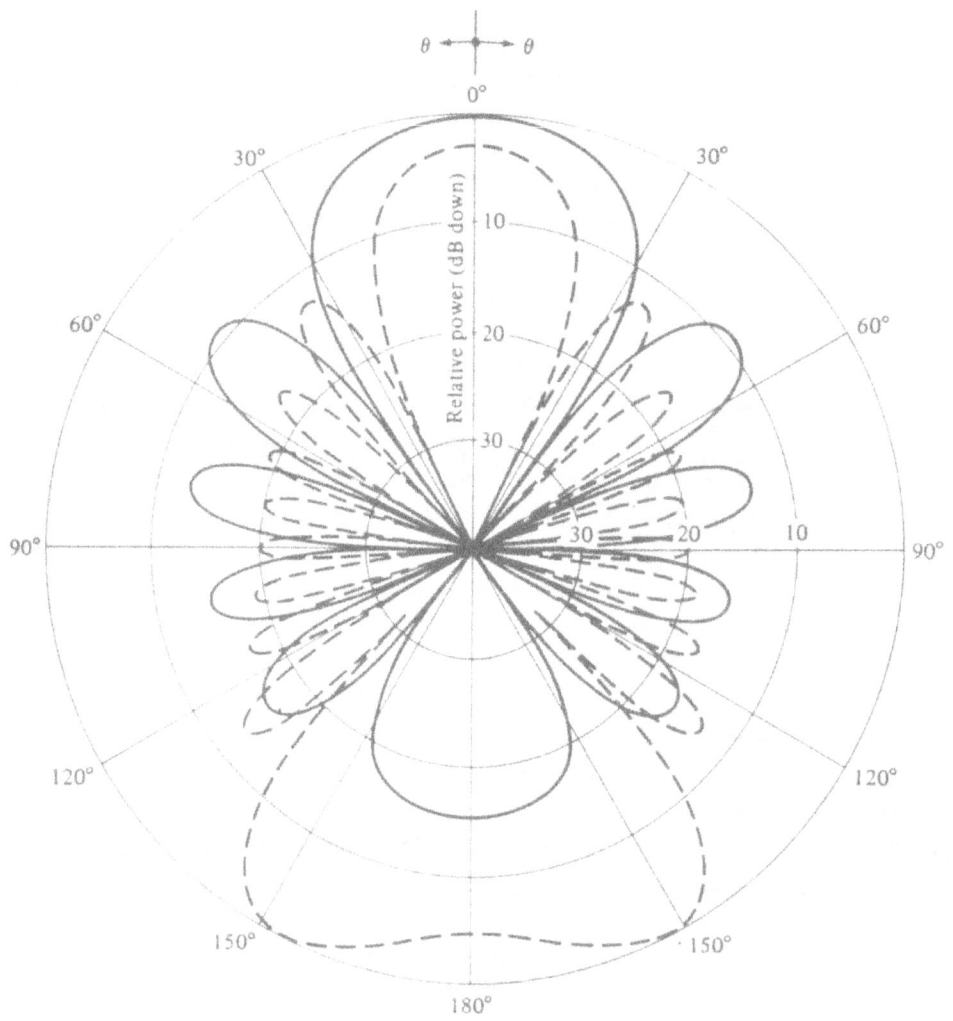

Solid line: $d = \lambda/4$
Dotted line: $d = \lambda/2$
$N = 10$
$\beta = -\left(kd + \dfrac{\pi}{N}\right)$

2. Directivity of a linear array
2.1. Directivity of a BSA

$$U(\theta) = |AF_n|^2 = \left[\frac{\sin\left(\frac{N}{2}kd\cos\theta\right)}{\frac{N}{2}kd\cos\theta}\right]^2 = \left[\frac{\sin Z}{Z}\right]^2 \quad (16.24)$$

$$D_0 = 4\pi \frac{U_0}{P_{rad}} = \frac{U_0}{U_{av}} \quad (16.25)$$

where:

$$U_{av} = \frac{P_{rad}}{4\pi}$$

The radiation intensity in the direction of maximum radiation $\theta = \pi/2$ in terms of AF_n is unity:

$$U_0 = U_{max} = U(\theta = \pi/2) = 1$$

$$\Rightarrow D_0 = \frac{1}{U_{av}} \quad (16.26)$$

The radiation intensity averaged over all directions is calculated as:

$$U_{av} = \frac{1}{4\pi}\int_0^{2\pi}\int_0^{\pi}\frac{\sin^2 Z}{Z^2}\sin\theta\, d\theta\, d\phi = \frac{1}{2}\int_0^{\pi}\left|\frac{\sin\left(\frac{N}{2}kd\cos\theta\right)}{\frac{N}{2}kd\cos\theta}\right|^2 \sin\theta\, d\theta$$

Change variable:

$$Z = \frac{N}{2}kd\cos\theta \Rightarrow dZ = -\frac{N}{2}kd\cos\theta\, d\theta \quad (16.27)$$

$$U_{av} = -\frac{1}{2}\int_0^{\pi}\left[\frac{\sin Z}{Z}\right]^2 d\cos\theta\, d\theta = -\frac{1}{2N}\frac{2}{kd}\int_{\frac{Nkd}{2}}^{-\frac{Nkd}{2}}\left(\frac{\sin Z}{Z}\right)^2 dZ \quad (16.28)$$

$$U_{av} = \frac{1}{Nkd} \int_{-\frac{Nkd}{2}}^{\frac{Nkd}{2}} \left(\frac{\sin Z}{Z}\right)^2 dZ \qquad (16.29)$$

The function, $\left(\frac{\sin Z}{Z}\right)^2$, is a relatively fast decaying function as Z increases. That is why, for large arrays, where $Nkd/2$ is big enough (≥ 20), the integral (16.29) can be approximated by:

$$U_{av} \simeq \frac{1}{Nkd} \int_{-\infty}^{\infty} \left(\frac{\sin Z}{Z}\right)^2 dZ = \frac{\pi}{Nkd} \qquad (16.30)$$

$$D_0 = \frac{1}{U_{av}} \simeq \frac{Nkd}{\pi} \simeq 2N\left(\frac{d}{\lambda}\right) \qquad (16.31)$$

Substituting the length of the array $L = (N-1)d$ in (16.31) yields:

$$D_0 = 2\underbrace{\left(1 + \frac{L}{d}\right)}_{N}\left(\frac{d}{\lambda}\right) \qquad (16.32)$$

For a large array $(L \gg d)$:

$$D_0 \simeq 2L/\lambda \qquad (16.33)$$

2.1 Directivity of ordinary EFA

Consider an EFA with maximum radiation at $\theta = 0°$, i.e. $\beta = -kd$.

$$U(\theta) = |AF_n|^2 = \left\{\frac{\sin\left[\frac{N}{2}kd(\cos\theta - 1)\right]}{\left[\frac{N}{2}kd(\cos\theta - 1)\right]}\right\}^2 = \left(\frac{\sin Z}{Z}\right)^2 \qquad (16.34)$$

where: $Z = \frac{N}{2}kd(\cos\theta - 1)$.

$$U_{av} = \frac{P_{rad}}{4\pi} = \frac{1}{4\pi}\int_0^{2\pi}\int_0^{\pi}\left(\frac{\sin Z}{Z}\right)^2 \sin\theta d\theta d\phi = \frac{1}{2}\int_0^{\pi}\left(\frac{\sin Z}{Z}\right)^2 \sin\theta d\theta$$

Again, change of variables is used:

$$Z = \frac{N}{2}kd\cos\theta \Rightarrow dZ = -\frac{N}{2}kd\cos\theta d\theta \quad (16.35)$$

$$U_{av} = -\frac{1}{2}\int_0^{\pi}\left(\frac{\sin Z}{Z}\right)^2 d\cos\theta = -\frac{1}{2}\frac{2}{Nkd}\int_0^{-Nkd}\left(\frac{\sin Z}{Z}\right)^2 dZ$$

$$U_{av} = \frac{1}{Nkd}\int_0^{Nkd}\left(\frac{\sin Z}{Z}\right)^2 dZ \quad (16.36)$$

If (Nkd) is sufficiently large, the above integral can be approximated as:

$$U_{av} = \frac{1}{Nkd}\int_0^{\infty}\left(\frac{\sin Z}{Z}\right)^2 dZ = \frac{1}{Nkd}\frac{\pi}{2} \quad (16.37)$$

$$\Rightarrow D_0 \simeq \frac{1}{U_{av}} = \frac{2Nkd}{\pi} = 4N\left(\frac{d}{\lambda}\right) \quad (16.38)$$

It is seen that the directivity of an EFA is approximately twice as large as the directivity of the BSA (compare (16.38) and (16.31)). Another (equivalent) expression can be derived for D_0 in terms of the array length $L = (N-1)d$:

$$D_0 = 4\left(1+\frac{L}{d}\right)\left(\frac{d}{\lambda}\right) \quad (16.39)$$

For large arrays, the following approximation holds:
$$D_0 = 4L/\lambda \quad \text{if} \quad L \gg d \quad (16.40)$$

2.2 Directivity of HW EFA

If the radiation has its maximum at $\theta = 0°$, then the minimum of U_{av} was obtained as (16.12):

$$U_{av}^{min} = \frac{1}{2k}\frac{2}{Nd}\left[\frac{Z_{min}}{\sin Z_{min}}\right]^2\left[\frac{\pi}{2} + \frac{\cos(2Z_{min})-1}{2Z_{min}} + \text{Si}(2Z_{min})\right] \quad (16.41)$$

where:

$$Z_{min} = -1.47 \simeq -\frac{\pi}{2}$$

$$\Rightarrow U_{av}^{min} = \frac{1}{Nkd}\left(\frac{\pi}{2}\right)^2\left[\frac{\pi}{2} + \frac{2}{\pi} - 1.8515\right] = \frac{0.878}{Nkd} \quad (16.42)$$

$$D_0 = \frac{1}{U_{av}^{min}} = \frac{Nkd}{0.878} = 1.789\left[4N\left(\frac{d}{\lambda}\right)\right] \quad (16.43)$$

From (16.43), one can see that using HW conditions leads to improvement of the directivity of the EFA with a factor of 1.789. Equation (16.43) can be expressed via the length L of the array as:

$$D_0 = 1.789\left[4\left(1 + \frac{L}{d}\right)\left(\frac{d}{\lambda}\right)\right] = 1.789\left[4\left(\frac{L}{\lambda}\right)\right] \quad (16.44)$$

Example: Given a linear uniform array of N isotropic elements ($N=10$), find the directivity D_0 if:
 a) $\beta = 0$ (BSA)
 b) $\beta = -kd$ (Ordinary EFA)
 c) $\beta = -kd - \frac{\pi}{N}$ (Hansen-Woodyard EFA)

In all cases, $d = \lambda/4$.

a) BSA

$$D_0 \simeq 2N\left(\frac{d}{\lambda}\right) = 5 \quad (6.999 \text{ dB})$$

b) Ordinary EFA

$$D_0 \simeq 4N\left(\frac{d}{\lambda}\right) = 10 \ (10 \text{ dB})$$

c) HW EFA

$$D_0 \simeq 1.789\left[4N\left(\frac{d}{\lambda}\right)\right] = 17.89 \ (12.53 \text{ dB})$$

3. Pattern characteristics of linear uniform arrays - recapitulation
   A. Broad-side array

NULLS ($AF_n = 0$):

$$\theta_n = \arccos\left(\pm\frac{n}{N}\frac{\lambda}{d}\right), \text{ where } n = 1,2,3,4,\ldots \text{ and } n \neq N, 2N, 3N,.$$

MAXIMA ($AF_n = 1$):

$$\theta_n = \arccos\left(\pm\frac{m\lambda}{d}\right), \text{ where } m = 0,1,2,3,\ldots$$

HALF-POWER POINTS:

$$\theta_h \simeq \arccos\left(\pm\frac{1.391\lambda}{\pi Nd}\right), \text{ where } \frac{\pi d}{\lambda} \ll 1$$

HALF-POWER BEAMWIDTH:

$$\Delta\theta_h = 2\left[\frac{\pi}{2} - \arccos\left(\frac{1.391\lambda}{\pi Nd}\right)\right], \ \frac{\pi d}{\lambda} \ll 1$$

MINOR LOBE MAXIMA:

$$\theta_s \simeq \arccos\left[\pm\frac{\lambda}{2d}\left(\frac{2s+1}{N}\right)\right], \text{ where } s = 1,2,3,\ldots \text{ and } \frac{\pi d}{\lambda} \ll 1$$

FIRST-NULL BEAMWIDTH (FNBW):

$$\Delta\theta_n = 2\left[\frac{\pi}{2} - \arccos\left(\frac{\lambda}{Nd}\right)\right]$$

FIRST SIDE LOBE BEAMWIDH (FSLBW):

$$\Delta\theta_s = 2\left[\frac{\pi}{2} - \arccos\left(\frac{3\lambda}{2Nd}\right)\right], \quad \frac{\pi d}{\lambda} \ll 1$$

B. Ordinary end-fire array

NULLS ($AF_n = 0$):

$$\theta_n = \arccos\left(1 - \frac{n}{N}\frac{\lambda}{d}\right), \text{ where } n = 1,2,3,\ldots \text{ and } n \neq N, 2N, 3N,$$

MAXIMA ($AF_n = 1$):

$$\theta_n = \arccos\left(1 - \frac{m\lambda}{d}\right), \text{ where } m = 0,1,2,3,\ldots$$

HALF-POWER POINTS:

$$\theta_h = \arccos\left(1 - \frac{1.391\lambda}{\pi Nd}\right), \text{ where } \frac{\pi d}{\lambda} \ll 1$$

HALF-POWER BEAMWIDTH:

$$\Delta\theta_h = 2\arccos\left(1 - \frac{1.391\lambda}{\pi Nd}\right), \quad \frac{\pi d}{\lambda} \ll 1$$

MINOR LOBE MAXIMA:

$$\theta_s = \arccos\left[1 - \frac{(2s+1)\lambda}{2Nd}\right], \text{ where } s = 1,2,3,\ldots \text{ and } \frac{\pi d}{\lambda} \ll 1$$

FIRST-NULL BEAMWIDTH:
$$\Delta\theta_n = 2\arccos\left(1 - \frac{\lambda}{Nd}\right)$$

FIRST SIDE LOBE BEAMWIDH:
$$\Delta\theta_s = 2\arccos\left(1 - \frac{3\lambda}{2Nd}\right), \quad \frac{\pi d}{\lambda} \ll 1$$

C. Hansen-Woodyard end-fire array

NULLS:
$$\theta_n = \arccos\left[1 + (1-2n)\frac{\lambda}{2Nd}\right], \text{ where } n = 1, 2, \ldots \text{ and } n \neq N, 2N, \ldots$$

MINOR LOBE MAXIMA:
$$\theta_s = \arccos\left(1 - \frac{s\lambda}{Nd}\right), \text{ where } s = 1, 2, 3, \ldots \text{ and } \frac{\pi d}{\lambda} \ll 1$$

SECONDARY MAXIMA:
$$\theta_m = \arccos\left\{1 + [1-(2m+1)]\frac{\lambda}{2Nd}\right\}, \text{ where } m = 1, 2, \ldots \text{ and } \frac{\pi d}{\lambda} \ll 1$$

HALF-POWER POINTS:
$$\theta_h = \arccos\left(1 - 0.1398\frac{\lambda}{Nd}\right), \text{ where } \frac{\pi d}{\lambda} \ll 1, \ N\text{-large}$$

HALF-POWER BEAMWIDTH:
$$\Delta\theta_h = 2\arccos\left(1 - 0.1398\frac{\lambda}{Nd}\right), \text{ where } \frac{\pi d}{\lambda} \ll 1, \ N\text{-Large}$$

FIRST-NULL BEAMWIDTH:

$$\Delta\theta_n = 2\arccos\left(1 - \frac{\lambda}{2Nd}\right)$$

4. **3-D characteristics of a linear array**

In the previous considerations, it was always assumed that the linear-array elements are located along the z-axis, thus, creating a problem, symmetrical around the z-axis. If the array axis has an arbitrary orientation, the array factor can be expressed as:

$$AF = \sum_{n=1}^{N} a_n e^{j(n-1)(kd\cos\gamma+\beta)} = \sum_{n=1}^{N} a_n e^{j(n-1)\psi}, \quad (16.45)$$

where a_n is the excitation amplitude and $\psi = kd\cos\gamma + \beta$.

The angle γ is subtended between the array axis and the radius-vector to the observation point. Thus, if the array axis is along the unit vector $\hat{a}$:

$$\hat{a} = \sin\theta_a \cos\phi_a \hat{x} + \sin\theta_a \sin\phi_a \hat{y} + \cos\theta_a \hat{z} \quad (16.46)$$

and the radius – vector to the observation point is:

$$\hat{r} = \sin\theta\cos\phi\hat{x} + \sin\theta\sin\phi\hat{y} + \cos\theta\hat{z} \quad (16.47)$$

the angle γ can be found from:

$$\cos\gamma = \hat{a}\cdot\hat{r}$$
$$= \sin\theta\cos\phi\sin\theta_a\cos\phi_a\hat{x} + \sin\theta\sin\phi\sin\theta_a\sin\phi_a\hat{y} + \cos\theta\cos\theta_a\hat{z}$$
$$\Rightarrow \cos\gamma = \sin\theta\sin\theta_a\cos(\phi-\phi_a) + \cos\theta\cos\theta_a \quad (16.48)$$

If $\hat{a} = \hat{z}$ ($\theta_a = 0°$), then $\cos\gamma = \cos\theta$, $\gamma = \theta$.

www.antentop.org

44 Feet Dipole. Where is the Truth?

By: Arnie Coro, CO2KK, CUBA

After more than half a century working with professional and amateur radio antenna systems... All I can say this is a never ending subject!! The 44 feet long dipole, 22 feet on each leg came out of the work of the prematurely deceased Len Cebik W4RNL... (**Figure 1** shows the 44-Feet Dipole Antenna.)

Len was an antenna modelling fanatic and guru at the same time who had the wisdom to share his gifted knowledge with everyone who approached him, like yours truly. When I asked Len about him choosing the 44 and 88 feet dipoles, he answered with both e'mail and printed correspondence.

By spending endless hours with sophisticated antenna modelling programs that are a real challenge to learn how to use because of the steep learning curve, Len found out that for the frequency range of the amateur bands between 7 and 29 megaHertz, the 44 feet dipole, fed with open wire transmission line of between 300 and 400 ohms impedance was an excellent compromise, using, of course,

Arnie Coro, CO2KK

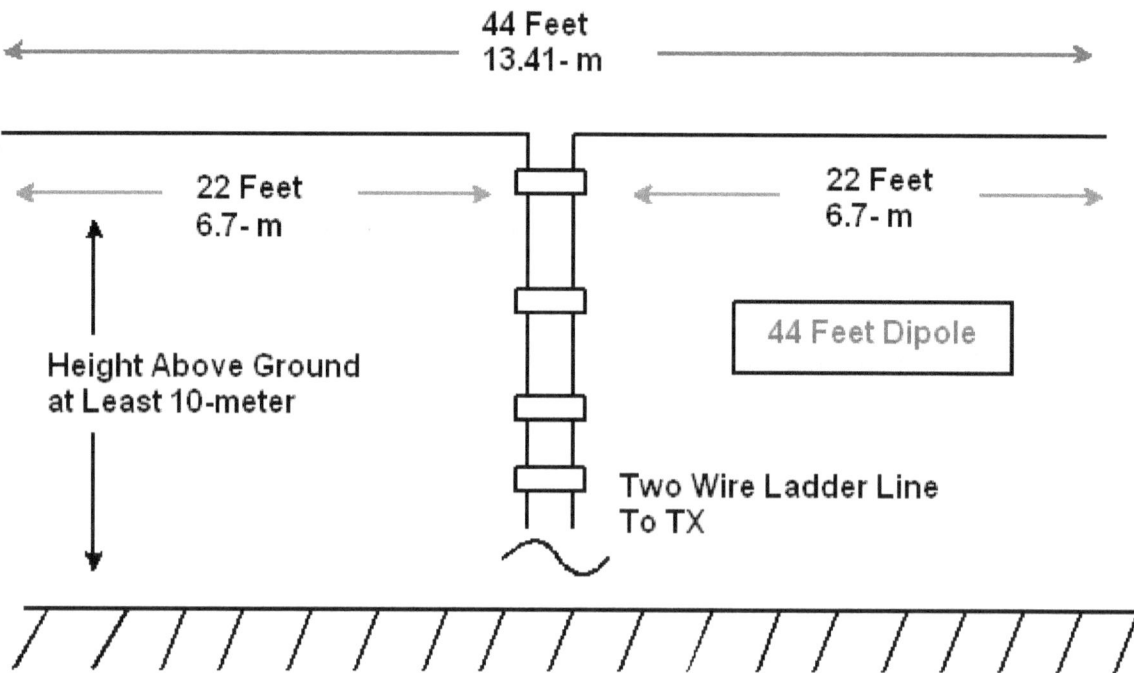

Figure 1
44 Feet Dipole Antenna

a well designed and well built antenna tuner plus a good quality SWR meter. 44 feet, when converted to meters will give you a length of 13.41 meters overall, or 6.70 meters for each leg. Likewise, the 88 feet overall length is 26.82 meters of 13.41 meters on each leg...

(**Figure 2** shows the 88- Feet Dipole Antenna.)

Note I.G.: *It is very, very sophisticated and old question – what optimal length for a wave ham antenna is. We may found lots different antennas and lots different way to match the antenna with a feed line. However, let's simulate the input impedance of the antenna with help MMANA. Below you can see SWR (at 350- Ohm Terminal) for 44 feet Antenna placed at 10 and 20- meters above real (simulated by MMANA of course) ground. As you can see the 44- feet antenna really has at the amateur bands SWR/Impedance that could be matched with simple ATU.*

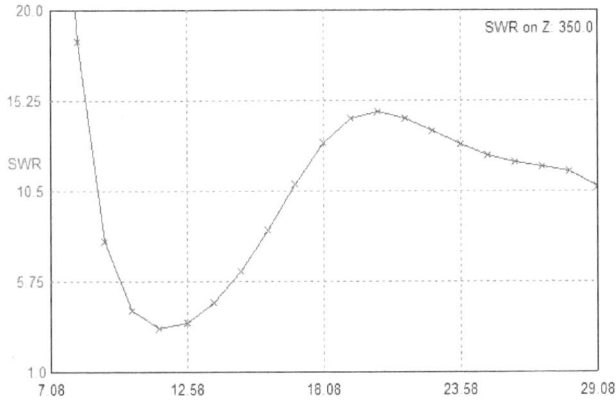

SWR (at 360-Ohm Terminal) of a 44-Feet Antenna Placed at 10- meters above Real Ground

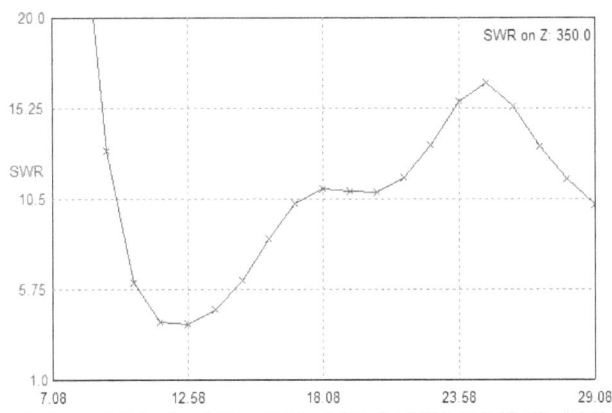

SWR (at 360-Ohm Terminal) of a 44-Feet Antenna Placed at 20- meters above Real Ground

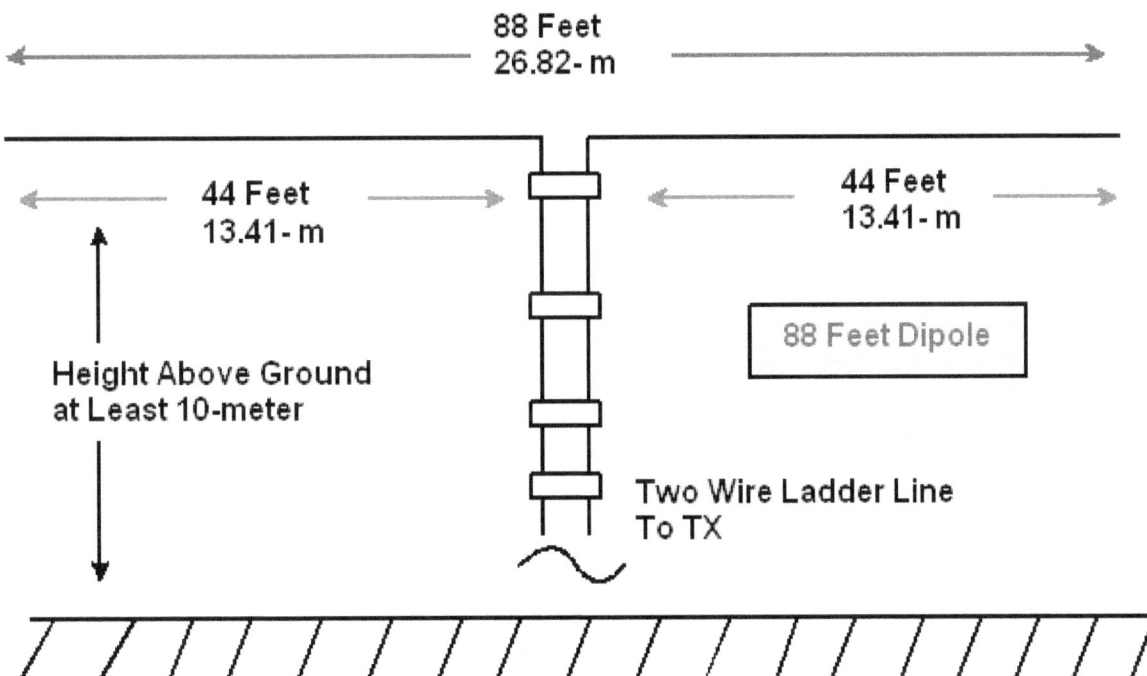

Figure 2
88- Feet Dipole Antenna

Note I.G.: *Below you can see SWR (at 350- Ohm Terminal) for 88- Feet Dipole Antenna placed at 10 and 20- meters above real (simulated by MMANA) ground. As you can see the 88- feet antenna really has at the amateur bands SWR/Impedance that could be matched with simple ATU.*

As you can see, the 88- Feet Dipole Antenna is much better matched with 350- Ohm feed line in compare to 44-Feet Dipole Antenna. It is only proved the old amateur's wisdom on antenna- the bigger the better. On my look the 88-Feet Antenna is more preferable, but room for this one!

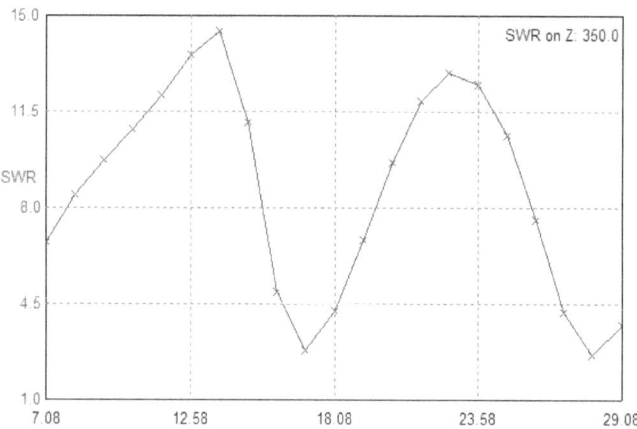

SWR (at 360-Ohm Terminal) of a 88-Feet Antenna Placed at 10- meters above Real Ground

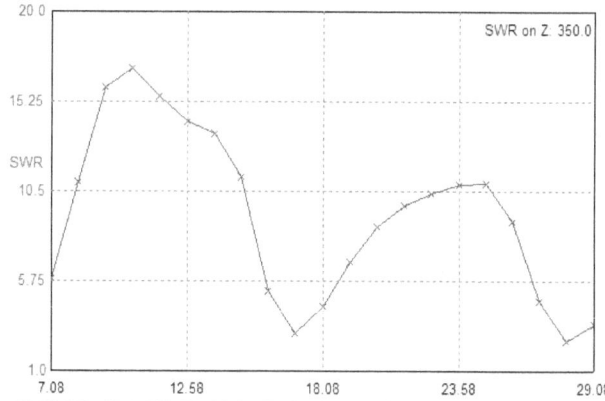

SWR (at 360-Ohm Terminal) of a 88-Feet Antenna Placed at 10- meters above Real Ground

Now, using the standard accepted formula for calculating the "approximate" length of a wire dipole for the HF frequency range from about 5 to 25 megaHertz.... 143 / frequency in megahertz let us play with a little math amigo !!!

143/ 7.15 for the half wave resonance on the 40 meters band will yield exactly 20 meters, so 10 meters on each side or moving to Imperial British Measures 65.6 feet overall or 32.8 feet on each leg. (**Figure 3** shows the 66- Feet Dipole Antenna.)

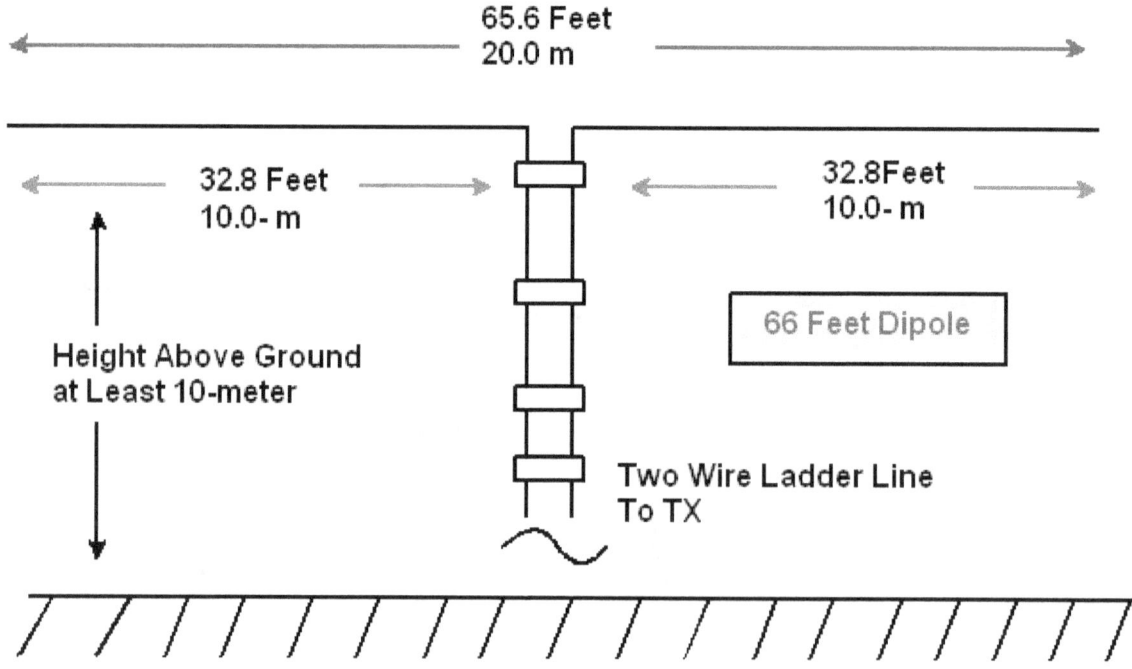

Figure 3
66- Feet Dipole Antenna

Note I.G.: *Below you can see SWR (at 350- Ohm Terminal) for 66- Feet Dipole Antenna placed at 10 and 20- meters above real (simulated by MMANA) ground. As you can see the 66- feet antenna really has at the amateur bands SWR/Impedance that could be matched with simple ATU.*

As you can see, the 66- Feet Dipole Antenna is better matched with 350- Ohm feed line in compare to 44 and 88- -Feet Dipole Antenna. However, if we take in consideration the Gain and Diagram Directivity, the 88- Feet antenna will beat the 66 and 44-Feet Dipole Antenna.

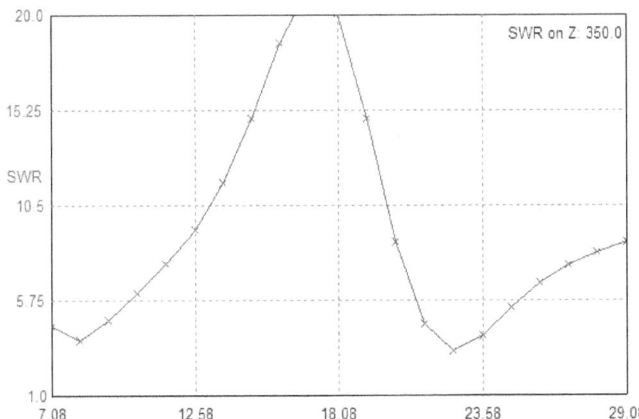

SWR (at 360-Ohm Terminal) 66-Feet Antenna Placed at 10- meters above Real Ground

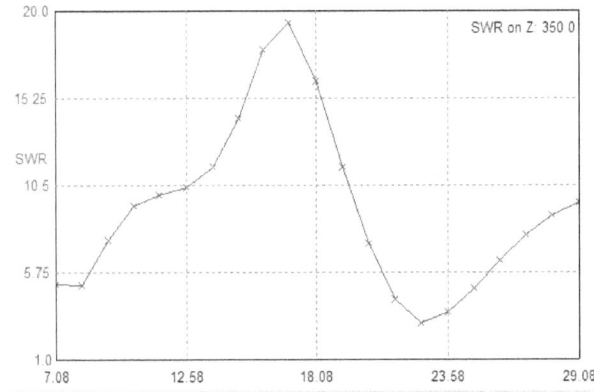

SWR (at 360-Ohm Terminal) 66-Feet Antenna Placed at 20- meters above Real Ground

So Len Cebik's 44 feet dipole falls short to achieve true half wave resonance on 40 meters of course... But because he recommended to use the open wire transmission line, preferably 400 to 500 ohms impedance, the open wire line will provide " the missing length" by means of the first few feet of the transmission line

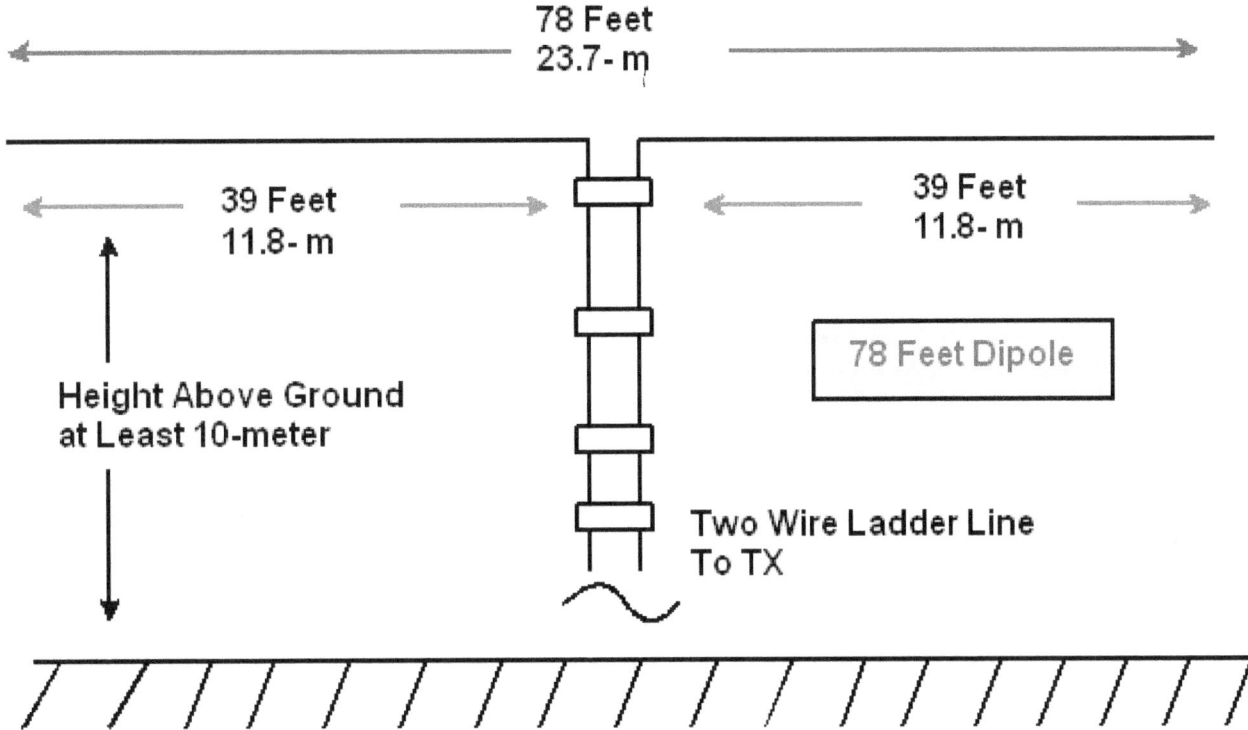

Figure 4
78- Feet Dipole Antenna

After doing my homework with the mathematics, I decided to use a different dipole length for my multiband HF wire antenna... The experimental antenna length that worked best for me was 39 feet long on each leg, or 78 feet overall length, that is 11.8 meters each leg or 23.7 meters overall length.

(**Figure 4** shows the 78- Feet Dipole Antenna.)

In order to make it fit into the available space the last 6 feet of the antenna on each leg are placed at a 90 degrees angle ...

Note I.G.: Below you can see SWR (at 350- Ohm Terminal) for 78- Feet Dipole Antenna placed at 10 and 20- meters above real (simulated by MMANA) ground. As you can see the 78- feet antenna really has at the amateur bands SWR/Impedance that could be matched with ATU.

The antenna is real compromise between 88- 66- and 44- Feet Dipole Antennas. Gain and Diagram Directivity of the antenna is not bad.

Files MMANA for the antennas you may download at

http://www.antentop.org/018/co2kk_018.html

73! va3znw

In order to increase the bandwidth of my 78 feet or 23.7 meters dipole fed with open wire line of 400 ohms, I used two wires on each leg, that are spread apart from the center insulator at a distance of 50 centimeters... The open wire transmission line is connected to a junction box, where I can select a 1 to 1 or a 4 to 1 balun that is connected via coaxial cable to the PI network antenna tuner...

73 and DX
Arnie Coro
radio amateur CO2KK
Host of Dxers Unlimited radio hobby program
Radio Havana Cuba

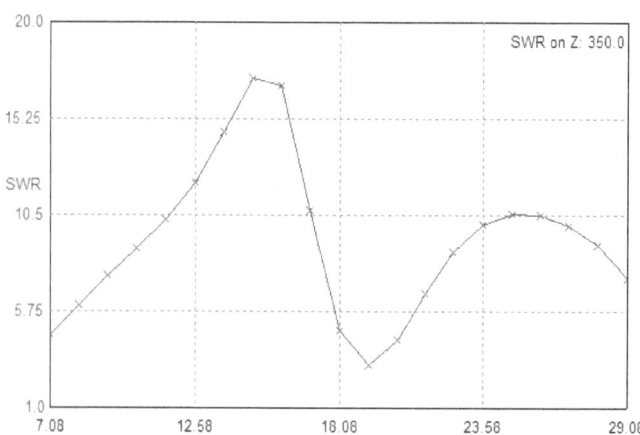

SWR (at 360-Ohm Terminal) 78- Feet Antenna Placed at 10- meters above Real Ground

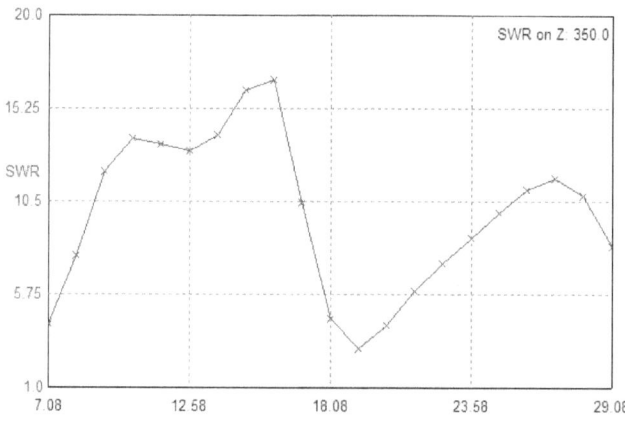

SWR (at 360-Ohm Terminal) 78- Feet Antenna Placed at 20- meters above Real Ground

Lawn Antenna

James R Kellner, K2MIJ, Hicksville, NY

Ok Gang in case you don't remember me....I am the guy who ran a couple of "Forks" as a dipole awhile back and actually worked W1AW/5 Oklahoma running 5 watts SSB from my trusty Yaesu FT-817....Well I have a new one for you and help will be appreciated to reach my goal!

I will digress a bit before making my help request.

Four days ago I received a text from my portable ops partner that 15 meters was showing some signs of opening, We had been discussing for awhile now at the lackluster band condx, I texted him back that I had noticed condx. seeming to be improving and that while cleaning the shack I found the "Fork Antenna" and had been giving it a good workout trying to get myself a spot on the RBN or possibly another contact with it.
He laughed and said "Knowing you, I am sure you will succeed" Standing in the yard while we were texting I noticed the two empty folding aluminum lawn chairs sitting there and I said to him "Perhaps I'll see if I can throw some RF into a couple of lawn chairs and see what happens!"

Well of course as soon as we finished our conversation I folded up the two chairs, stuck one under each arm and off to the shack I went. Five minutes later "The Lawn Chair Antenna" was born!

Running the 817 @ 5 watts into my favorite LDG Z11 AT, 6 feet of RG8X coax and a 4:1 balun the chairs loaded without a hitch on 10 thru 40! (40 you say....keep reading!)

But would it radiate? I threw out a quick "CQ CQ de K2MIJ K2MIJ" on 20 meters and was immediately spotted by N7TR Reno Nevada. Eureka, Success!

Could it actually work? IS DX possible?

I started working in earnest from that moment and the results have nothing less than amazing! First contact was with John N0JA in MO 20 meter SSB John gave me a 52 and I was ecstatic! Next up was Morris W4REX in FL a 53 report on 17 meters SSB and the mind blowing contact of the day was my 3rd QSO, Lacy HA3NU in Hungary on 17 meters SSB a typical 59 "contest exchange" report...BUT....I made it across the pond with 5 watts into a pair of lawn chairs sitting on a bed inside the shack!

Lawn Antenna

I was getting a complex, I could work phone contacts without a problem but a CW contact was an effort in futility!

Finally after a number of attempts....my 1st CW contact with K1GHL on 40 meters was a complete delight! The LDG Z11 tuner found a match quite easily on 40 meters but making a contact I thought would be at best a stroke of complete luck given the antennas obvious inefficiency at that low of a frequency...Boy was I ever wrong! My first CQ out of the box netted me an RBN spot by VE2WU with a SNR of 9db! As I was picking myself up off the floor it took me a minute to realize that was my call coming back through the speaker...K1GHL Glenn in upstate NY gave me a 559 report but disappointingly came back on the next go around saying he had lost me in the noise and bid me a 73...Crestfallen, I soon was to have my spirits lifted when immediately after Glenn, Art W2NRA called me and we completed a solid 5 to 6 turnover QSO and a good SKCC exchange! I now have 6 completed 2 way CW contacts on 40 meters. Furthest distance is New Hampshire with Randy KX1NH.... Yes the lawn chairs will work on 40! My 1st CW DX contact finally came via S.E. station AO8BWC Canary Islands on 20 meters...

In the past 4 days, CW and/or SSB, I have so far put 55 domestic and DX contacts in the log, 14 states NH, NJ, NY, FL, GA, AL, IN, MI, WI, IL, MO, AR, TX and CO and 9 countries Hungary, Canary Islands, Germany, Sweden, Ireland, Poland, Czech Republic, Wales and Russia!

I know, took a long time to get to the subject line help request but here goes, I started this completely as a lark but after seeing how well it can and will work...I am now looking to seriously attempt Worked All States...IF you would like to help me on my quest I would be happy to have you in my log either CW or SSB. I will post here on the list from time to time when and where the "Lawn Chairs" will be operating. I will also spot myself on the QRP spots site...TNX in advance!

Well Gang Summer is winding down and Fall is right around the corner... before you sadly drag those lawn chairs into the garage or shed for their long Winter nap, why not extend their life and tuck a couple under your arms and bring em into the shack instead! Don't worry about that the Wife or Kids will think your crazy, Trust me, Your a Ham...They already know it! ;-)

1st time using Dropbox so hopefully if I have done everything correctly you can see some pictures of the Lawn Chair antenna in it's operating position...also I included a couple of links to some youtube videos. One while I worked Jim NV9X IL in this past weekends SKCC Weekend Sprint(if you watch the video don't ask me why I pronounced Illinois as "Ill e noise", I have no clue!) and a SSB contact with HG7T in the WAE contest...

72/73, Jim K2MIJ

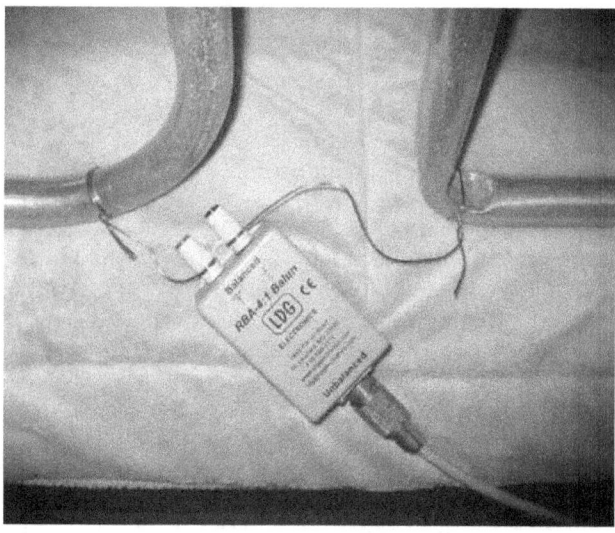

Feed Point of the Lawn Antenna

Link for videos with Lawn Antenna

HG7T SSB contact link-
http://youtu.be/ol6Or6RaWHM

NV9X SKCC CW link http://youtu.be/SjIKT10AaH4

Lawn Chair DXCC Countries/Entities worked list

Hungary - Canary Islands - Germany - Sweden - Ireland - Poland - Czech Rep. - Russia - Wales - Spain - Trinidad & Tobago - Greece - Slovenia - Belgium - Cuba - Puerto Rico - Canada - Bermuda - Croatia - Scotland - St. Lucia - England - Bonaire - Austria - Azores - Italy - Belarus -...more to follow!

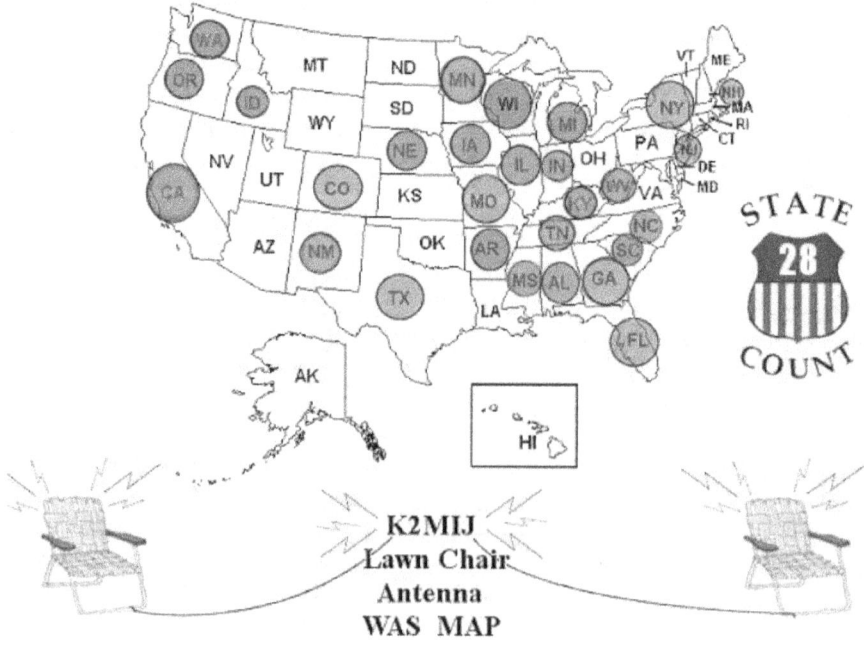

Buried Antennas for Emergency Communications

John J. Schultz W1DCG/W2EEY

Credit Line: 73 MAGAZINE, April 1967, pp.: 34- 35

In this article, W1DCG describes some of the properties of buried antennas, particularly in relation to their usefulness for amateur Civil Defense or emergency communications installations.

Many experiments have been conducted with sub-surface antennas in recent years to allow construction of bomb – proof communication sites and for communication with deeply submerged submarines.

Some scientists believe in the possibility of a super-conductive medium in the earth's crust so that antennas could be buried in the ground in an upside-down fashion and communications established using this "earth ionosphere" much the same as surface antennas work in conjunction with the ionosphere.

However, experimenters in this field have not been very successful and a buried antenna, for practical purposes, can be treated as having useful propagation only above the surface. The deeper the antenna is buried, the more inefficient it becomes because of the earth's absorption of the radiated energy.

What application do such antennas have for the amateurs? Few amateurs are faced with such a drastic situation that they can't put up some form of surface antenna, even if only attic antenna for short whip. However, for those engaged in Civil Defense or other emergency communications work, the installation of a buried, "back- up" antenna at a fixed station should be considered.

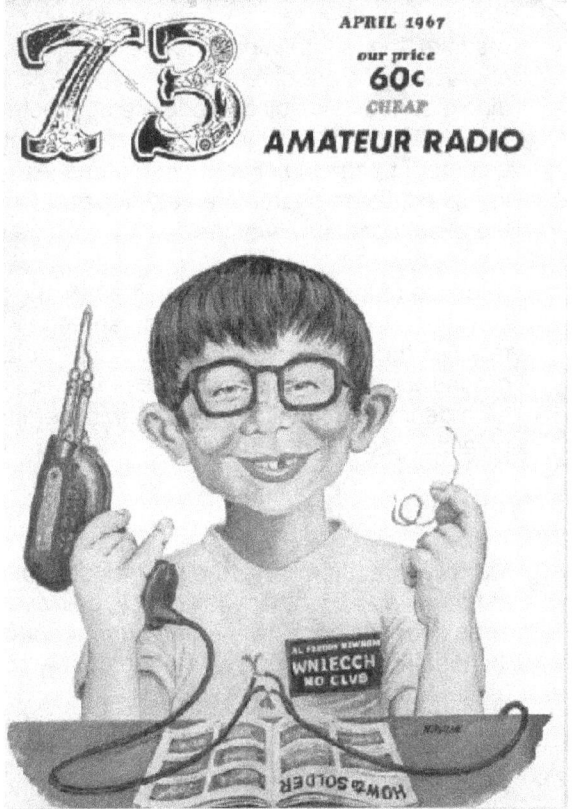

Front Cover: 73 MAGAZINE, April 1967

It is quite a contradiction to see so many times an emergency communications setup in a relatively protected area – the basement of some public building, for instance- and then to see the antennas on which the usefulness of the installation entirely depends, dangling loosely in the open liable to any extreme surface condition, natural or man- made.

Buried Antennas for Emergency Communications

The purpose of this article is to review some of the types of buried antennas which might be useful for amateur emergency communications and to present some of the results the author obtained with a buried 40- meter antenna.

Buried antenna properties

Because a buried antenna is immersed in a very lossy medium and because of the sudden difference in medium which, a radiated wave encounters at the interface of earth and air, a number of factors are drastically different for buried antennas as compared to an antenna in air.

Because of the antenna being in a different medium, the length/impedance versus frequency characteristics are different. For instance, a simple dipole buried 1' in soil of moderate conductivity would be about 17 ½ ' long for 20 meters and have a center impedance of about 450 ohms. As may be imagined, these figures are very dependent upon the exact conductivity of the soil.

Because of the interface between earth and air, the radiation from a horizontal, buried antenna when it reaches the surface sets up a vertically- polarized ground wave. This factor, of course, is ideal for emergency communications work with vertically – polarized mobile stations.

Antenna forms

Many different forms have been tried for buried antennas and even complex directive arrays have been constructed. However, for amateur purposes, the dipole and 100' long wire are probably most useful forms. (**Figure 1**)

The formula for the length of a dipole depends upon ground conductivity as well as other factors and would not be of much use to the average amateur.

The best procedure for constructing a dipole is simply to cut it to 80% of the free – space length and then trim the ends equally until the lowest SWR is achieved. If buried in an area where ground conditions remain stable, the length does not have to be changed again.
In areas where ground conditions and surface conditions (snow, extreme changes of vegetation) are not stable, the 100' long wire should be used.

Although an antenna coupler, such as a transmatch, is required to allow compensating for impedance changes with varying ground conditions, the antenna can then also be used for multiband operation. In typical soil the input resistance of such an antenna will vary from 50 to 600 ohms and the reactance from +- j400 ohms over the 2- 20 MHz range. The first resonance will be between 750 and 1800 kHz, which makes it effective from 80 meters on down. This type of antenna has been used by the Army in Viet Nam with good result over short tactical distances.

Construction

Whether a dipole or 100' long wire is used, the wire used for construction must be insulated along its length from the soil and care must be taken that moisture does not penetrate the tips of the wire or the connection to the feed line. Teflon insulated wire, numbers 22 to 26, is particularly suitable. Perhaps a less expensive method is to run plain rubber insulated wire inside plastic hosing. The ground connection for the 100' long wire can be a standard 4 or 5' TV type ground rod.

Efficiency

Many methods have been used for measuring the efficiency of buried antennas. Perhaps the most realistic for amateur purposes is to compare the field strength from a buried antenna to a good, surface quarter- wave vertical. Experiments made on this basis have showed buried antennas of the dipole and 100' long wire variety, when compared to a surface antenna resonant at the same frequency, to be about 40 db down for a burial depth of 1'. Roughly, this is about twice the order of magnitude reduction in signal strength as would take place between a 8' loaded 80- meter whip and a full – size quarter- wave 80- meter vertical.

Experimental results

The author constructed a 100' long –wire buried about 8" and operated on 40- meters. No impedance measurement were made but proper loading could be easily achieved with the use of a transmatch – type coupler, although some retuning was necessary periodically depending on whether the soil surface was moist or dry.

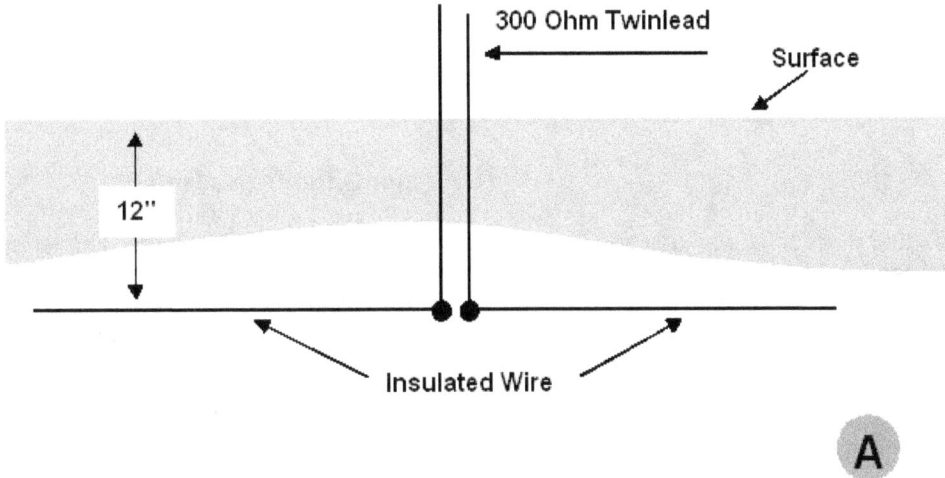

A

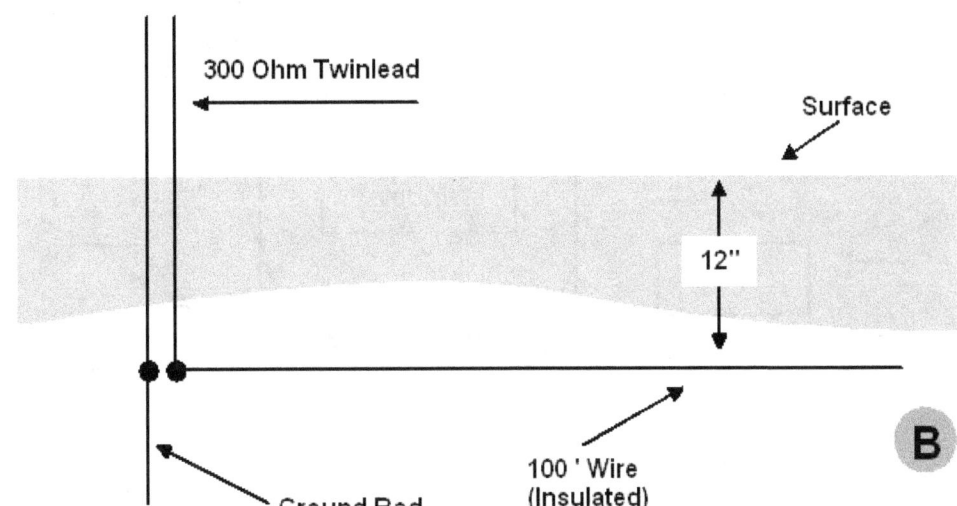

B

No surface, vertical 40 meter antenna was available to make signal comparisons but comparisons were made with a 40-meter dipole elevated about 40 feet. On local and short – skip contacts, the buried antenna was never better than 7 "S" units below the dipole with the average being 8-9 "S" units.

Conclusion

Buried antennas still offer many possibilities for experimentation. The main caution to observe is that the length, impedance and other parameters of surface antennas cannot be used.

Buried antennas are terribly inefficient as compared to almost any type of surface antenna except perhaps extremely short, unloaded whips.

But, for emergency communications installations, they do offer the possibility of having a standby antenna which is easily installed and which can be pre-tuned and immediately available for use should something happen to the installation's primary antenna.

.....W1DCG

Reference

For those who would like further detailed, engineering data on buried antennas, the following compilation of articles is extremely useful: IEEE Transactions, Vol. AP-11, May.1963. Special Issue on Electromagnetic Waves in the Earth. IEEE, Box A, Lenox Hill, New York 21, New York.

Simple Broadband Antenna for the 40- meter Band

Igor Grigorov, va3znw

The Simple Broadband Antenna was designed on base of my local environment and taking into account the ease/cheap to do. **Figure 1** shows the antenna. Three points there are need to install the antenna. There is one point at right fence second point at left fence and third point at a window on the second floor of my house. For antenna wire it was used electrical copper wire in diameter 18- AWG in strong black insulation (33-cent/m, Home Depot is supplier).

Antenna is simple to do. A 10- meters length of the wire is fastened by the ends to both fences. Leave 50- 80 cm free wire length at both sides at initial installation.

Figure 2 shows the wires. The length, from one side, should be used for tuning antenna in to the resonance, at second side should connect with feed line. Center of the wire is heightened and attached to the window frame.

Ground of the antenna included a wire going on to the fence down to the earth. Near the earth the wire was connected to an earth wire. The earth wire was made of three wires. One wire was galvanized bare wire in 18 AWG. Two wires (similar to the antenna wire) were attached to the stranded wire by ties. The triple wire had length near 3.5 meter.

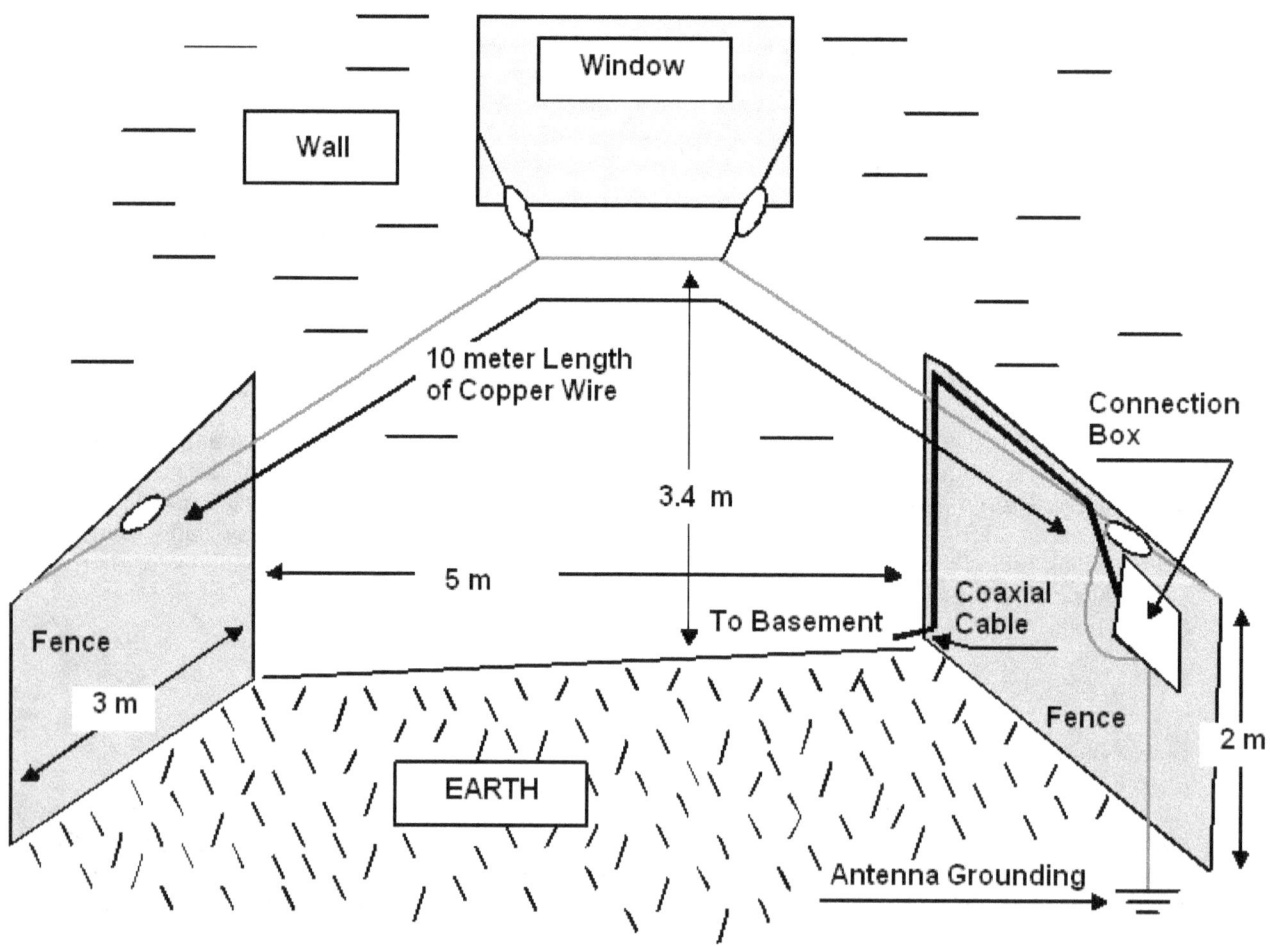

Figure 1 Simple Broadband Antenna for the 40- meter Band

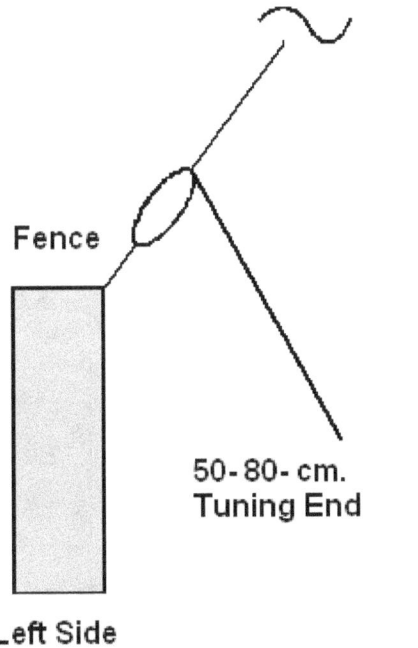
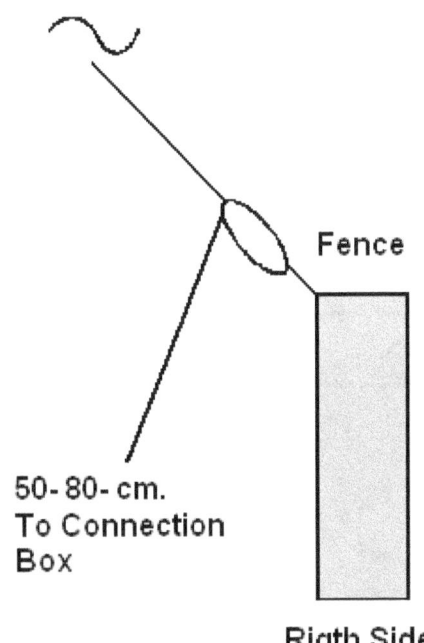

Figure 2 Free wires at initial installation of the antenna

The wire was dug in the earth at depth near 10 centimeters. The wire was connected to the Artificial Earth of the Helical Antenna (**Reference 1**). The antenna fed by a 50-Ohm coaxial cable. The cable was attached at upper side of the fence then went down to the ground and lead in to the basement through a ventilation hole. **Figure 3** shows schematic of the antenna.

Picture 1 shows Feed Point of the Simple Broadband Antenna for the 40- meter Band. **Picture 2** shows Center of the Simple Broadband Antenna for the 40- meter Band.

Picture 3 shows Tuning End of the Simple Broadband Antenna for the 40- meter Band. **Picture 4** shows Coaxial Cable going along the Fence. **Picture 5** shows Coaxial Cable going into my Basement. Take attention, there are three cables are going into the basement. One coaxial cable fed Helical Antenna (Reference 1) second coaxial cable fed the Simple Broadband Antenna for the 40- meter Band and the third coaxial cable used to for my experimental with antennas (in free from experimental time a 144/430- MHz antenna is connected to the cable).

Picture 1 Feed Point of the Simple Broadband Antenna for the 40- meter Band

Picture 2 Center of the Simple Broadband Antenna for the 40- meter Band

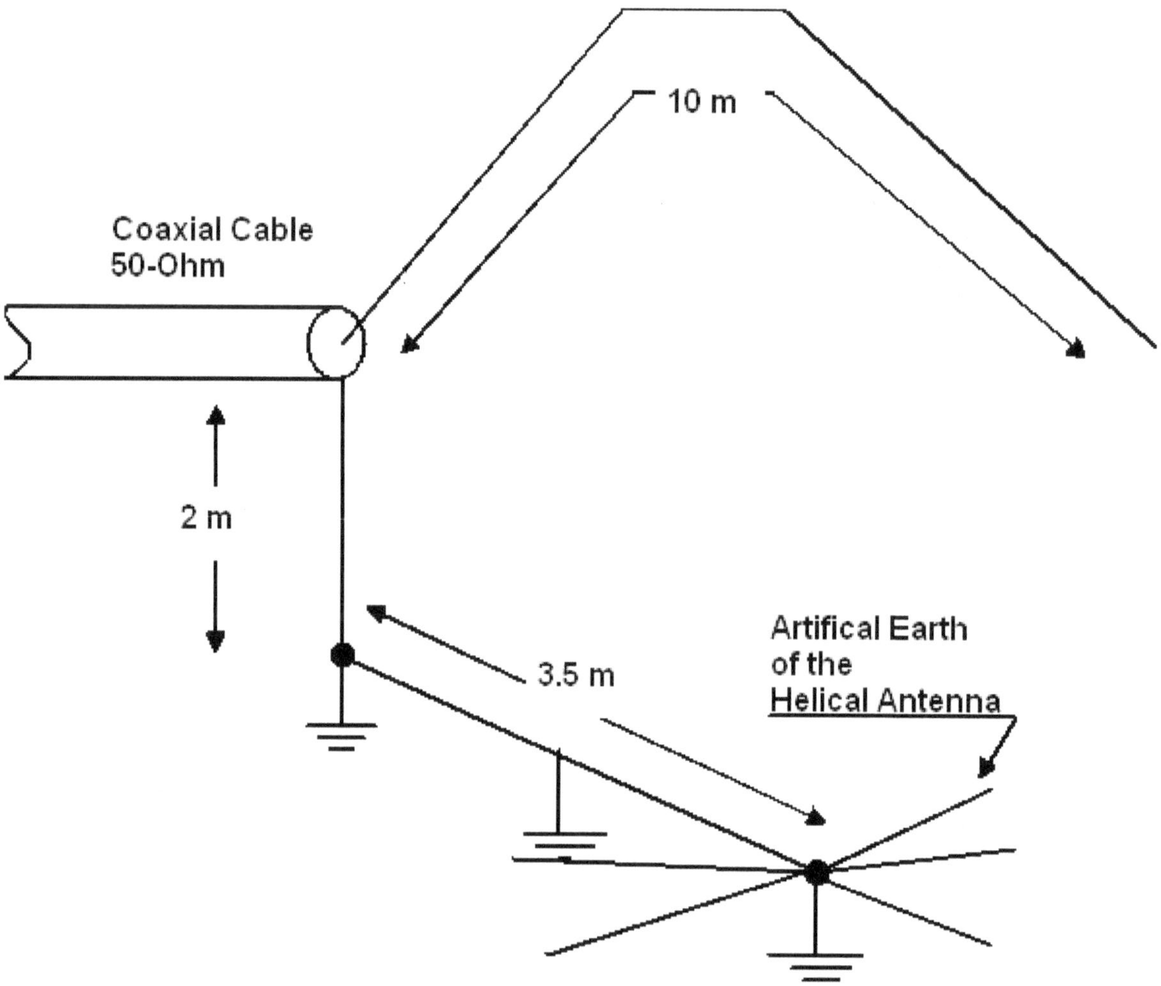

Figure 3 Schematic of the Simple Broadband Antenna for the 40- meter Band

Picture 3 Tuning End of the Simple Broadband Antenna for the 40- meter Band

Picture 4 Coaxial Cable going along the Fence

A connection box was installed at the feed point of the antenna. **Figure 4** shows drawing of the box. Box made of the base a plastic food container. An RF socket SO-239 was installed at the box.

Feeding coaxial cable was connected to the socket. Arrestor resistor 4k7/5-Wtt was soldered to the bridge of the socket.

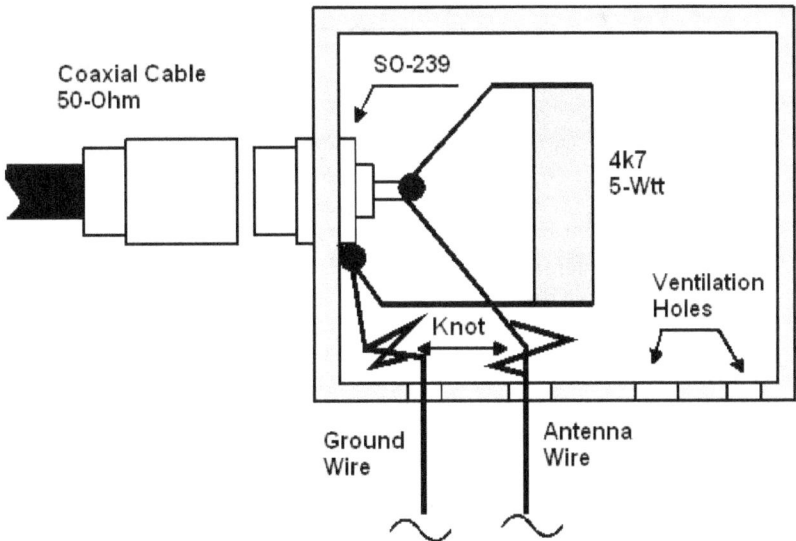

Figure 4 Connection box

The resistor leaks the static from antenna wire to the ground. Antenna and ground wire went into the box from the down side. The wires had knot on it inside the box. The knot did allow the wires go off the box. Two ventilation holes (for removing condensed water) were drilled at the down side of the box.

Connection Box was placed on the fence at South – East side. So the most part of the day the box was shined by the Sun. At the autumn I discovered that tabs at the cover had cracks at the bending. The tags went away from the cover when this one was opened. **Picture 6** shows the cover with cracked retaining tags. Take attention that one retaining tag had cutoff (the cutoff is near antenna socket). It was happened in 1.5 years from the antenna installation. However I had a spare cover to fix the box.

Picture 5 Coaxial Cable going into my Basement

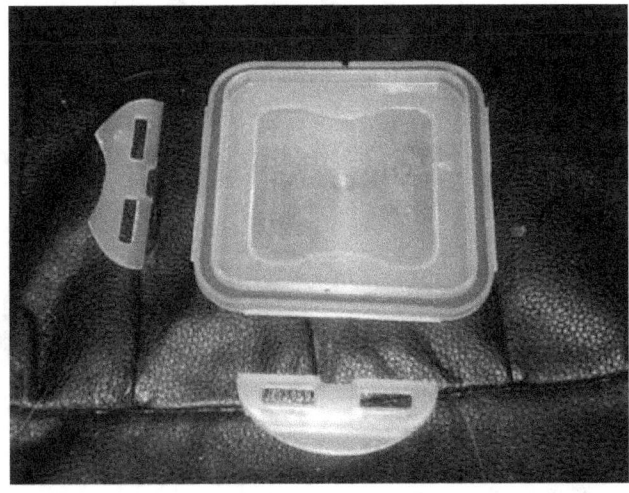

Picture 6 Cover with cracked retaining tags

Tuning the antenna to resonance is a simple matter. Connect an antenna analyzer (I use to a MFJ- 259B) to the antenna terminal (ground of the coaxial cable should be connected with antenna) or (that is better) to end of the coaxial cable going from the antenna. Measure the resonance frequency. The frequency would be lower the 7.0- MHz. Then do shortening of the antenna wire from the left side. I do not recommend cut the wire, just fold it like it is shown at **Figure 5**.

I got antenna impedance 75- 85- Ohm (at zero reactance) at 7.0- 7.350- kHz. The data is practically matched with data obtained from Antenna Simulator EZNEC for MMANA. The antenna should be simulated by this software. In spite of the antenna impedance is not strictly 50- Ohm at the 40- meter Band the antenna could match with my IC- 718 and K-1 without any ATU.

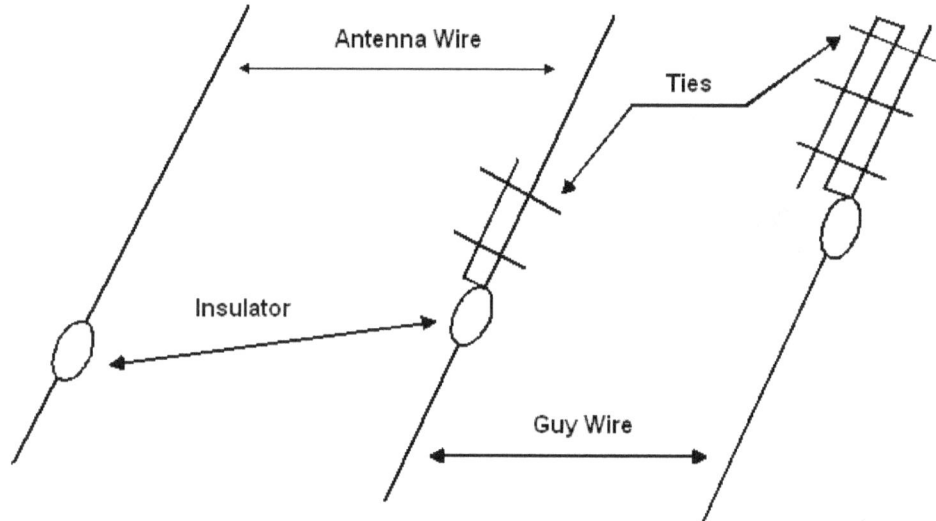

Figure 5 Folding Antenna Wire

At the 30- 10- meter Bands the antenna could match with transceiver with help of MFJ VERSA TUNER II or LDG Z-11 PRO. It is possible to install a simple ATU inside the connection box to get SWR 1.0: 1.0 at the 40- meter Band. However the 30- 10- meter band would be suffered at this case. It is possible to feed the antenna with TV- 75- Ohm coaxial cable. It is possible to buy very cheap such cable (ever intended for underground placement) at the E-Bay.

Antenna impedance and the resonance frequency of the antenna depend on the ground conditions. At wet ground the resonance goes down and input impedance goes down. At the dry (or frosty) ground the resonance goes down and input impedance goes up. So it is useful take control on the antenna at the season changes. However it is possible to adjust antenna length that the season changes do not hindered to the antenna resonance.

The input impedance of the antenna and the resonant frequency depends on the state of the earth. When moist earth resonance frequency of the antenna is reduced, the input impedance of the antenna is reduced. When dry land (or frozen), the resonant frequency of the antenna increases, and also increases its input impedance. This should be considered when setting up the antenna to resonance. It is useful from time to time, for example in rainy or very dry weather to control the resonant frequency of the antenna. This will adjust the antenna so that the Changes in the weather will not affect significantly changes its resonant frequency.

Reference

1. http://www.antentop.org/017/va3znw_017.htm

73/72!
va3znw

Photo 7 Antenna at Winter Time

Directional Antenna UA6AGW V.7.00

Aleksandr Grachev, UA6AGW

Credit Line: CQ-QRP # 43 (Summer, 2013) pp.: 21-27.

The antenna was born after numerous experiments that were made in past three years. Russian Patent # 125777 was obtained for the antenna. *Prototype* of the antenna is described in *Reference 1, 2*. Some experimenters on the born of the antenna are described in *Reference 3*. **Figure 1** shows the Directional Antenna. **Figure 1** shows all antenna dimension and placement of the antenna parts.

The antenna has some parts that are similar to the *prototype*. Loop part of the antenna made of a coaxial cable and this one is placed vertically. There are two phase- shift capacitors- C1 and C2. However horizontal wires have some modifications. Two long wires are connected to one part of the loop. One of the long wires (that is placed in the direction of the maxima radiation) form the main lobe. Other one long wire suppresses back radiation. Two short wires provide symmetrical of the antenna.

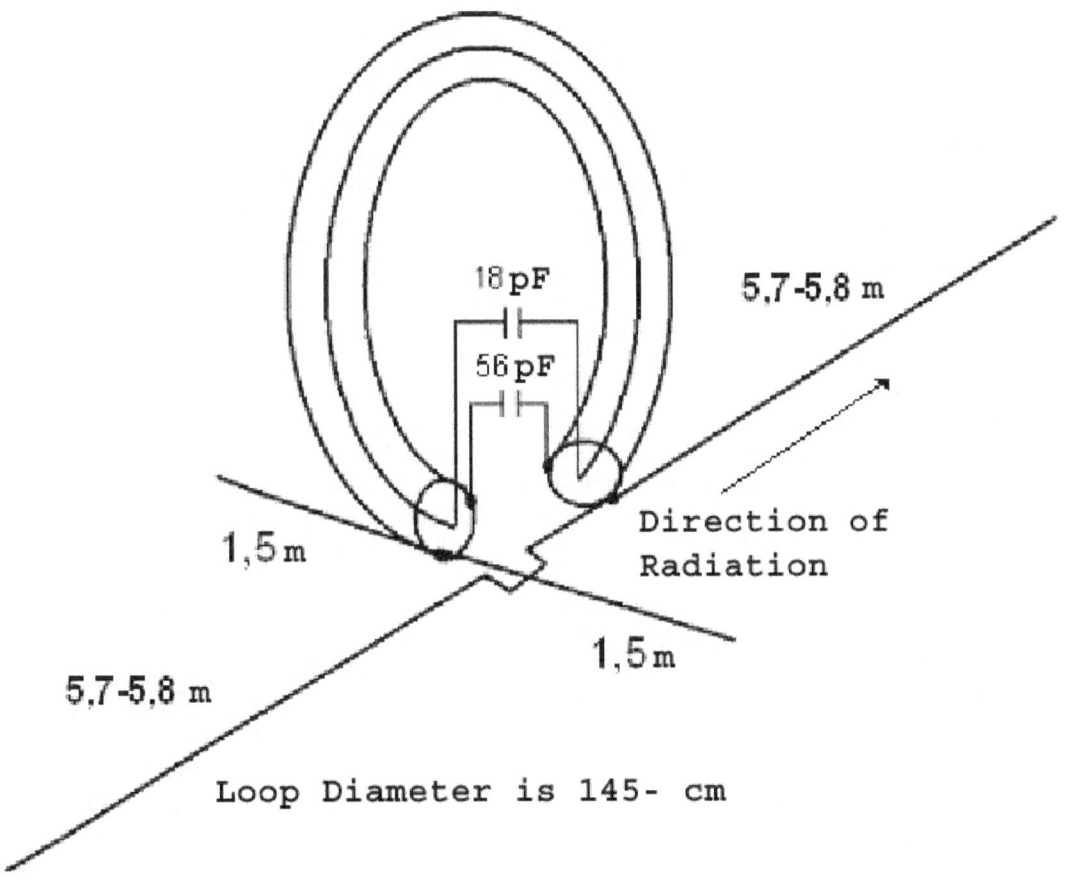

Figure 1 Directional Antenna UA6AGW V. 7.00

Design: Loop is made from so- called half- inch coaxial cable with crimped copper braid. The copper braid is covered by two layers of protection lacquer then covered by electrical protection plastic tube. It is made for weather – proof sustain. **Picture 1** shows the antenna. Usual plastic fishing poles are used for the form for long wires. Inside the fishing pole a multi-cored copper wire is going through. Thin ends of the fishing pole are changed by light aluminum wire in diameter 8- mm. The multi- cored wire is connected to the aluminum wire.

Short wires are placed along plastic tube in 14- mm diameter. The tube is not only support for the short wires. Rope guys going from the ends of the tube to ends of the fishing poles provided rigid of the antenna structure. **Picture 2** shows mounting of the plastic tube and wire montage of the horizontal wires.

Picture 1 Directional Antenna UA6AGW V. 7.00

Mast of the antenna has height in 8- meters. Two water tubes made the mast. First tube that is sitting on the ground is a metal tube in 48- mm OD. It is 5- meter long. The second one, that holds the antenna structure, is plastic tube in 42- mm OD. The plastic tube is 3- meter long. The plastic tube is inserted inside the metal tube that allows rotate the antenna. A simple home- made adaptor (made of from two pieces of metal water tubes in diameter 48- and 55- mm) is used for connection the mast's tubes. **Picture 3** shows jointing of the plastic and etal tube.

The antenna fed by a coupling loop. For simplicity of the design the coupling loop made from the feeding coaxial cable. **Figure 2** shows the coupling loop before it is circulated to loop. Length of the coaxial cable to be used for the coupling loop is 200- mm. Plastic from the length of the coaxial cable is removed on to 10- mm in the center and from two ends. Then braid of the coaxial cable is removed at the center. Inner conductor is soldered to the braid at the far (right) end of the length.

Направленная антенна UA6AGW v. 7.00

Александр Грачёв UA6AGW

Heading of the Article

Picture 2 Mounting of the Plastic Tube and Wire Montage of the Horizontal Wires

Picture 3 Jointing Mast's Tubes

Then the cable is turned to loop. Far end of the length is soldered to the first (left) side of the prepared cable. (In Russia the method of the making the coupling loop sometimes is named -method of the DF9IV-)

The coupling loop is fastened to the upper part of the antenna's loop with help of a Scotch and ties. Below there are several simple rules how to install the coupling loop.

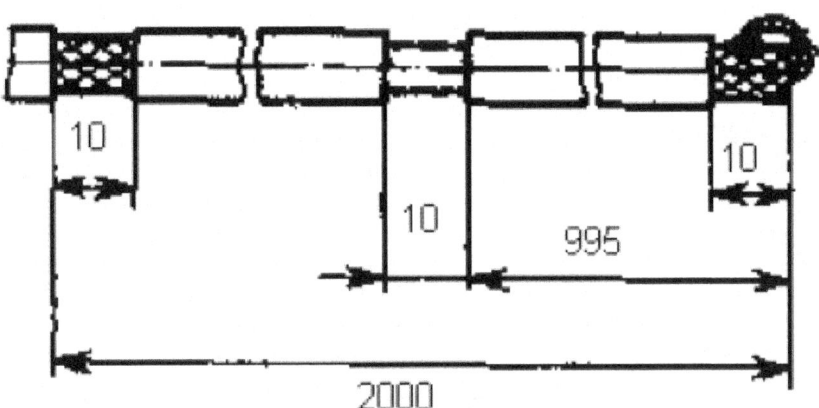

Figure 2 Preparation of the Coupling Loop for the Directional Antenna UA6AGW V. 7.00

At first, find on the antenna loop a point that is equidistance from left and right side of the C2. It is *the point of symmetry* of the antenna.

At second, find the point of symmetry of the coupling loop. The coupling loop is mounted in the top of the antenna loop. Point of symmetry of the coupling loop should concur with the point of symmetry of the antenna. Picture 4 shows the coupling loop on the antenna.

At third, to fasten with help of the cable ties the coupling loop to the antenna loop at the distance of 6-8- cm from the point of symmetry of the antenna loop.

Antenna was tuned (when it was placed on mast) in height 5- meters (it is from the ground to the top of the mast). Horizontal wires and matching box with capacitors was at 3.5- meters above the ground. A 2- meter ladder was used by me for tuning the antenna. Antenna works fine ever at the small height. F/B ratio was near 20-dB in this case. Antenna was tuned to 7080- kHz in mind that the resonance frequency move up (to 7100- kHz) at the height 8- meter. Antenna is simple to tune to the resonance. It may be tuned with help C2 (56- pF at Figure 1) to maximum RF- voltage at the long horizontal wire or with help receiver to maximum receiving signal.

However I used fixed capacitors at my antenna. So I did tuning of the antenna by changing length of the horizontal wires.

Preliminary Summary: All good features of the prototype antennas (**References 1, 2**) are not lost at the Antenna UA6AGW V. 7.00. The features are: small dimension, easy to do, interference immunity, ability to operate at small height. Antenna has input impedance 50- Ohm. Antenna bandwidth (at SWR 2.0:1.0, measured by an antenna analyzer) is near 150- kHz. **Picture 5** shows scan from the antenna analyzer.

Within ten month Antenna UA6AGW V. 7.00 was tested in the Air. The antenna was compared with Antenna UA6AGW V.40.02 (**Reference 4**). **Picture 6** shows screen shot a SDR –Transceiver. Left picture shows the transceiver with antenna UA6AGW V.40.02 right picture show screen shot with Antenna UA6AGW V. 7.00.

Picture 4 Coupling loop on the Directional Antenna UA6AGW V. 7.00

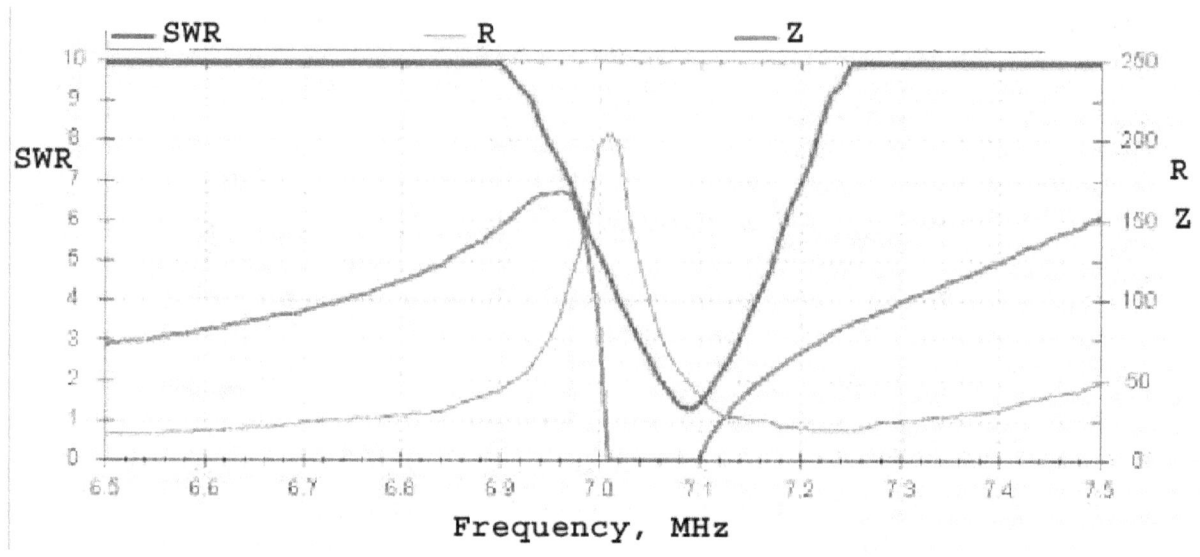

Picture 5 Data for the Antenna UA6AGW V. 7.00

Antennas were switched through small period of time. Antenna UA6AGW V. 7.00 has the same strength of the reception in main lobe (compare to Antenna UA6AGW V.40.02) however stations going not from main lobe are weak in reception.

Figure 3 shows DD of the Antenna UA6AGW V. 7.00 in horizontal plane. The DD was obtained by measuring of the signal of control transmitter located at distance 1- km from the antenna.

Width of the DD in horizontal plane (at level 3- dB) is near 60 degree. Level F/B is not less 20- dB. Level F/S is near 15 dB. At day time the antenna suppresses local stations (in radius 300- 350- km) on to 20- 30- dB. So the main lobe in vertical plane should have angle in 35- 40 degree.

Antenna gain was estimated according to diagram and formulas from page 61 at **Reference 5**. Antenna gain for the antenna should be near 10- dB. **Figure 4** shows the diagram.

Summary:

1. Antenna UA6AGW V. 7.00 differs little from the prototype but turns to directional antenna.
2. Having small dimension the antenna compare to them has good F/B ratio in 20-dB.
3. F/B and F/S ratio for the antenna is constant at working range.
4. There is a real possibility to design small directional antenna for the 80- meter Band.
5. Antenna keeps all merits of the prototype.
6. Antenna is simple to design and easy to tune

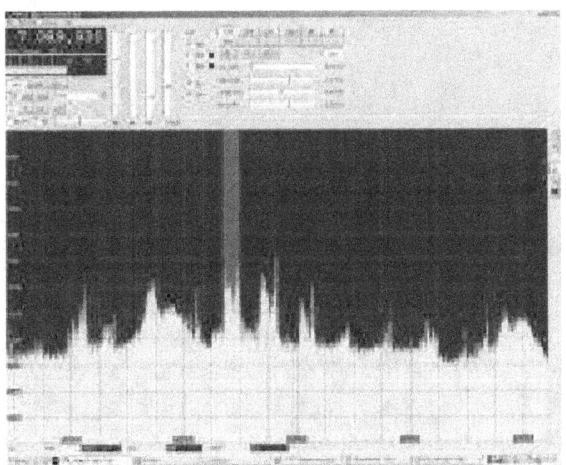
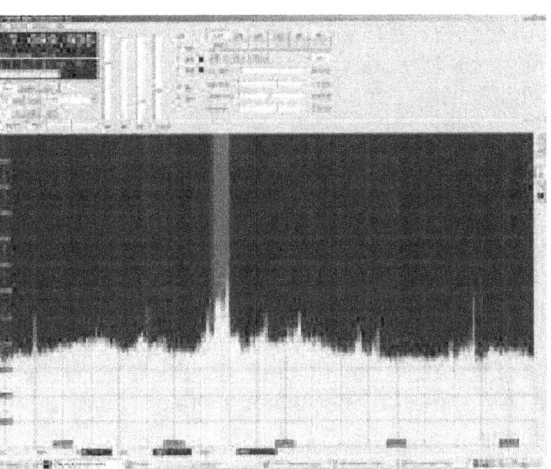

Picture 6 Screen Shot from SDR Transceiver with Compared Antennas

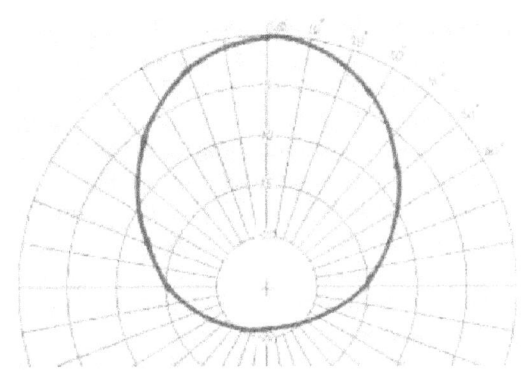

Figure 3 DD of the Antenna UA6AGW V. 7.00 in Horizontal Plane

References:

1. Antenna UA6AGW V.40, Aleksandr Grachev, Radio # 2, 2011, pp.: 59- 61
2. Antenna UA6AGW V. 80, Aleksandr Grachev, Radio 8, 2011, pp.: 60- 61.
3. Experimenters with Magnetic Loop Antennas, Aleksandr Grachev, CQ-QRP # 27, 2009, pp.: 9- 11.
4. http://www.antentop.org/017/ua6agw_md_017.htm
5. Antennas, Karl Rothammel, Nash Gorod, Minsk- 2001

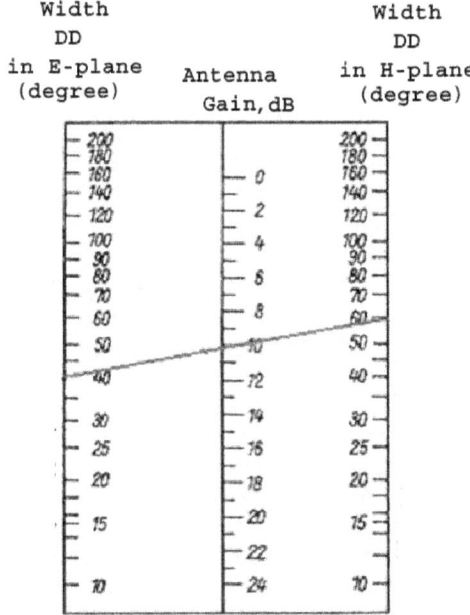

Figure 4 Diagram to Estimate Antenna Gain for the Antenna UA6AGW V. 7.00

Antenna UA6AGW in Experimenters by RU1OZ

Nikolay Chabanov, RU1OZ

Credit Line: CQ- QRP # 42. Spring- 2013. Pp.: 15-17.

In my native city Archangelsk my house is located very close to the Power Transmission Line. High noise from the line forced me to use Magnetic Loop Antennas. The antennas could work very effective and may eliminate the electrical noise. I can make QSOs with EU and JA (at 10- 18- MHz) using only 4-watts with the Magnetic Loop Antenna. But... Magnetic Loop Antenna has some disadvantage for me. It is a narrow pass-band (I need tune the Loop across the amateur band). My variable capacitor at the Loop is sparked at power close to 5-watts (that is way I used to only 4-watts with the Magnetic Loop).

In one of the lucky day I have read article about the UA6AGW Antenna. The antenna straight away attractive my attention because:

1. Antenna takes small room. So I may place it instead my Magnetic Loop Antenna
2. Pass Band of the antenna is 150... 200- kHz. So, I do not need retune the antenna inside the working Band.
3. The horizontal wires are lowered the RF-Voltage across the variable capacitor. That should be no sparking at the variable capacitor.

The three above mentioned factors were main point to make the antenna. For making the antenna I used a length of a 75-Ohm coaxial cable in diameter 13- mm. To turn the antenna in rigid design the length of coaxial cable was hid inside a plastic Hula- Hoop that had diameter 80- cm. **Figure 1** shows the schematic of the antenna for the 10 and 14- MHz.

Picture 1 Antenna UA6AGW for 14 and 18- MHz Installed at my Room

I used usual variable capacitor (12- 495- pF) from an old radio as C2. Usual variable capacitor 10- 70-pF was used as C1. Horizontal wires were made from a multicore copper wire in diameter 3- mm. Antenna for testing purpose was installed inside of my room. **Picture 1** shows the antenna in my room. Antenna was tested at 18 and 14- MHz Bands. The length of the horizontal wires for the bands was 2.5- meter.

Straight away I noticed that in comparison with old Magnetic Loop (made from the same coaxial cable) the antenna UA6AGW worked better on to reception. And it is no sparking at the variable capacitor at 5-watts RF-power. I used to an old surplus military radio R-143 that fed by 12- Volts. At the voltage the radio gives no more the 5-watt output power.

Антенна UA6AGW — рамочно-лучевая (v.30/20. 00)

Николай Чабанов RU1OZ

Heading of the Article

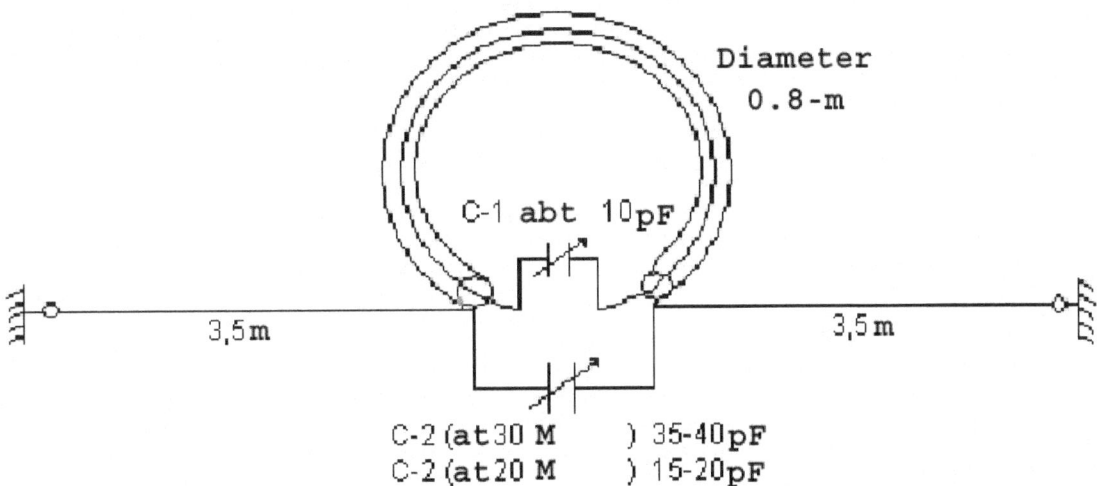

Figure 1 Antenna UA6AGW

Antenna worked well at the 18 and 14- MHz Bands. SWR was 1.2: 1.0 at the Bands. Pass Band was (at SWR 2.0: 1.0) 150- kHz at both bands. I made several CW and PSK- 31 QSOs with EU, Russia and Ukraine. As usual reports were 599 at the both ends.

Then I installed the antenna at my attic. The antenna was intended for 14 and 10-MHz. Horizontal wires were 3.5- meters in length in this case. Pass Band at 14- MHz Band was 230- kHz at SWR 2.0:1.0. **Picture 2** shows the antenna in my attic. Coupling loop for the antenna was made from copper wire in diameter 4- mm. Length of the wire was equal to the diameter of the antenna UA6AGW (80- cm in my case). Coaxial cable RG-58 (length 18- meter) was going from the antenna to my radio. Antenna was tuned to needed band manually by C2.

Antenna in attic had height near 8- 9- meter above the ground. Antenna worked well at all- seasons- at dry summer, at wet autumn and in the winter when snow blanket in 30- 50- cm thick was placed on the roof above the antenna.

Later a LW-Antenna in 41- meter length and at 7- meter height above the ground was installed at my location. The antenna was fed from the end by Fuchs method (see **Reference 1**). The LW was compared with Antenna UA6AGW. Program WSPR (that showed levels of the receiving stations on the computer screen) objectively proved the antenna UA6AGW advantage with the 41- meter length of wire antenna.

Picture 2 Antenna UA6AGW in the Attic

Reference 1
Josef Fuchs (OE1JF) Antenna: Patent Description
http://www.antentop.org/016/oe1jf_016.htm

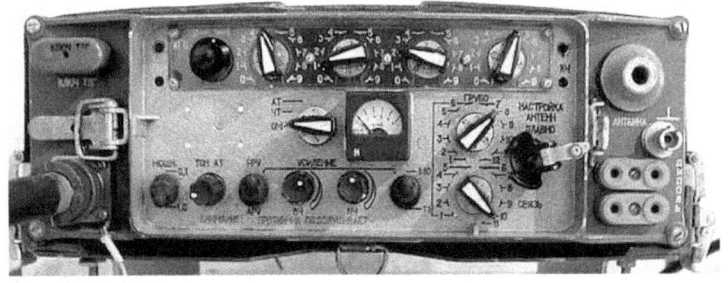

Ex- Soviet Military Radio R-143

Antenna UA6AGW V.40.20

Aleksandr Grachev, UA6AGW

The experimental antenna made on the base of previously described versions of the Antenna UA6AGW for 40- meter Band. Aim of the experiment was to reduce the room that the horizontal wire takes. For the purpose one of the horizontal wires was bended. **Figure 1** shows schematic of the antenna.

Antenna was tuned (by C2) to the 40- meter Band and connected to my transceiver. It was found that the antenna lost its symmetrical properties. Electrical noise appeared at receiving mode. At transmitting an RF- Voltage appeared at the transceiver cabinet. However it was possible corrected (return symmetrical properties back) by lengthening on 10 percent the bended horizontal wire.

Diagram Directivity was taken for the antenna. DD of the antenna looks like an ellipse. Receiving from the sides of the ellipse is lower on 3- 6- dB. Receiving from the bended horizontal wire is lower on to 10- dB.

Summary: It was made simple, easy to do and tune antenna with directional properties.

73! UA6AGW

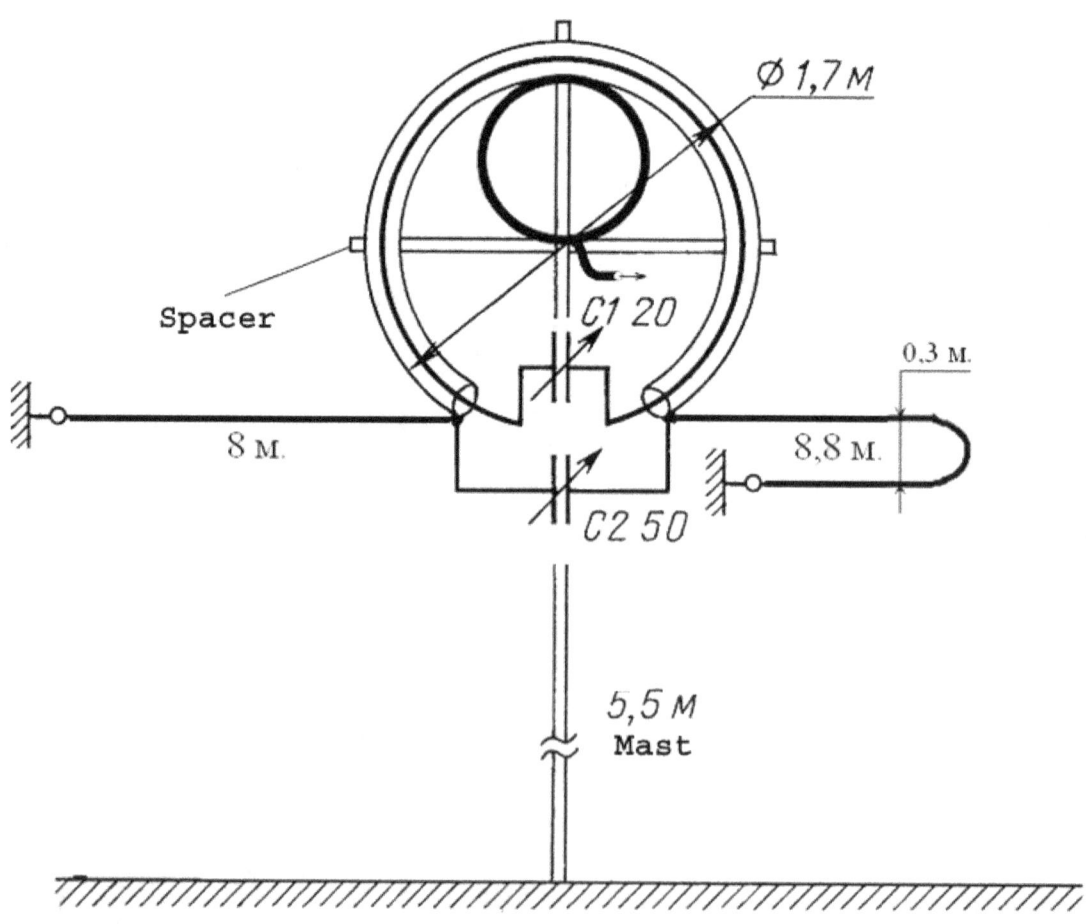

Figure 1 Antenna UA6AGW V.40.20

Field Antenna UA6AGW V. 40.21

Aleksandr Grachev, UA6AGW
Credit Line: CQ-QRP # 46 (Spring, 2014) pp.: 15- 19.

The antenna was designed for installation in a field conditions or limited space. Antenna may be installed at a low- height mast. Antenna does not required guys and takes small room for installation. **Figure 1** shows schematic of the antenna.

At first experimenters the two horizontal wires were bended (prototype Antenna UA6AGW V.40. 20, *ANTENTOP 01, 2014, p.: 42*). Then to reduce the occupied room the loop of the antenna was curved. **Picture 1** shows the Loop.

Two traverses for horizontal wires made from plastic fishing poles in 4- meter length of each. The horizontal wires made from audio cord in 1-mm diameter (18- AWG). **Picture 2** shows the traverse. Classical antenna insulators do not use in the antenna design. Fishing cord and plastic ties are used instead of those ones. The hook made from thick bare wire is installed at some ends of the fishing cord. Capacitor C1 is a high- voltage capacitor. Variable capacitor C2 is usual tuning capacitor 12- 495- pF from an old tube receiver.

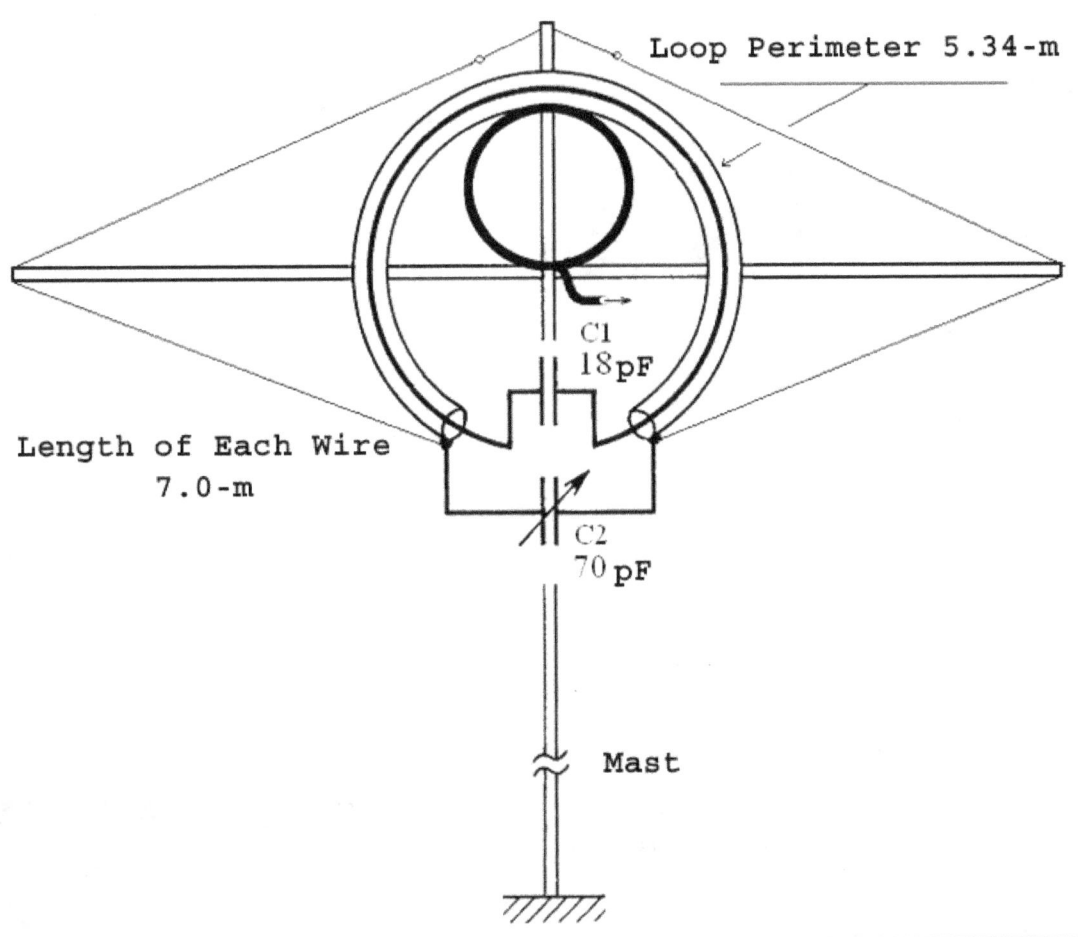

Figure 1 Schematic of the Field Antenna UA6AGW V.40.21

However every next plates of the capacitor are removed (to increase the working voltage). The capacitor connected to the loop only by two stator sections

Picture 3 shows ends of the fishing cord. Picture 4 shows connection box of the antenna. Picture 5 shows horizontal wires at the connection box.

Picture 1 Loop of the Field Antenna UA6AGW V.40.21

Picture 2 Traverse of the Field Antenna UA6AGW V.40.21

Picture 3 Ends of the Fishing Cord

Figure 2 shows the coupling loop before it is circulated to loop. Length of the coaxial cable to be used for the coupling loop is 200- mm. Plastic from the length of the coaxial cable is removed on to 10- mm in the center and from two ends. Then braid of the coaxial cable is removed at the center. Inner conductor is soldered to the braid at the far (right) end of the length. Then the cable is turned to loop. Far end of the length is soldered to the first (left) side of the prepared cable. The coupling loop is fastened to the upper part of the antenna's loop with help of a Scotch and ties. Below there are several simple rules how to install the coupling loop.

At first, find on the antenna loop a point that is equidistance from left and right side of the C2. It is *the point of symmetry* of the antenna.

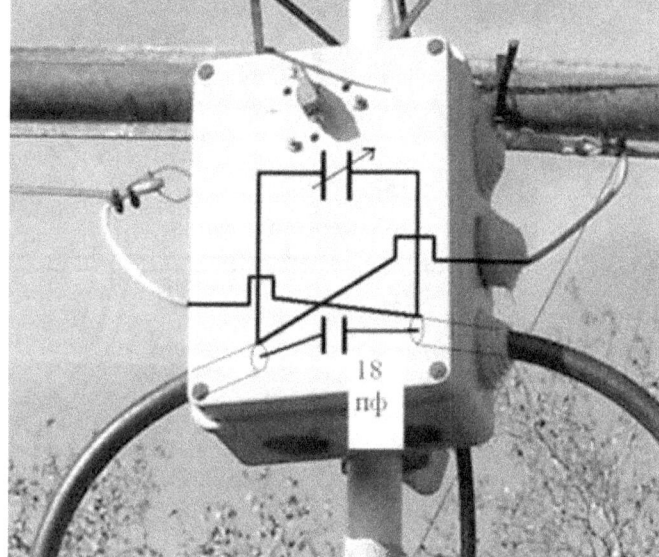

Picture 4 Connection box

Picture 5 Horizontal Wires at Connection Box

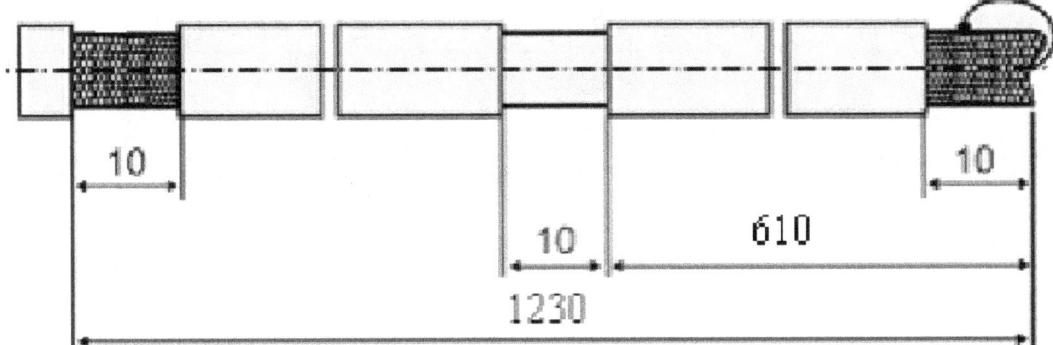

Figure 2 Preparation of the Coupling Loop

At second, find the point of symmetry of the coupling loop. The coupling loop is mounted in the top of the antenna loop. Point of symmetry of the coupling loop should concur with the point of symmetry of the antenna.

At third, to fasten with help of the cable ties the coupling loop to the antenna loop at the distance of 6-8- cm from the point of symmetry of the antenna loop.

Picture 6 shows the Field Antenna UA6AGW V.40.21. Antenna UA6AGW V.7.01 (horizontal wires are down) is seen on the background.

Antenna tuned to the resonance by capacitor C2. In receiving mode the antenna is tuned by maxima reception. In transmitting mode the antenna is tuned by maxima RF- Voltage at a horizontal wire. **Picture 7** shows parameters of the antenna measured by Antenna Analyzer AA-330M. Pass Band of the antenna at SWR 2.0: 1.0 is 90- kHz. However in the field conditions when capacitor C2 is accessible the antenna may be easy to retune.

Field Antenna UA6AGW V.40.21 was tested in the Air at height 6 and 4- meters above the ground. The antenna was compared with Antenna UA6AGW v.40.02 (**Reference 1**) installed on 7- meter mast. Antenna UA6AGW V.40.21 was tuned to 7120- kHz. Antenna UA6AGW V.40.02 was tuned to 7110- kHz.

Picture 6 Field Antenna UA6AGW V.40.21

Test at 6-meters height.

Antenna UA6AGW V.40.21 is low noise compare to Antenna UA6AGW v.40.02. **Picture 8** shows screen shot SDR transceiver with Antenna UA6AGW V.40.02. Receiving signal is near minus 120- dB. Ratio S/N is near 10- dB. **Picture 9** shows screen shot SDR transceiver with Antenna UA6AGW V.40.21. Receiving signal is near minus 120- dB. Ratio S/N is near 15- dB. The picture is taken at day light time with minimal time tag.

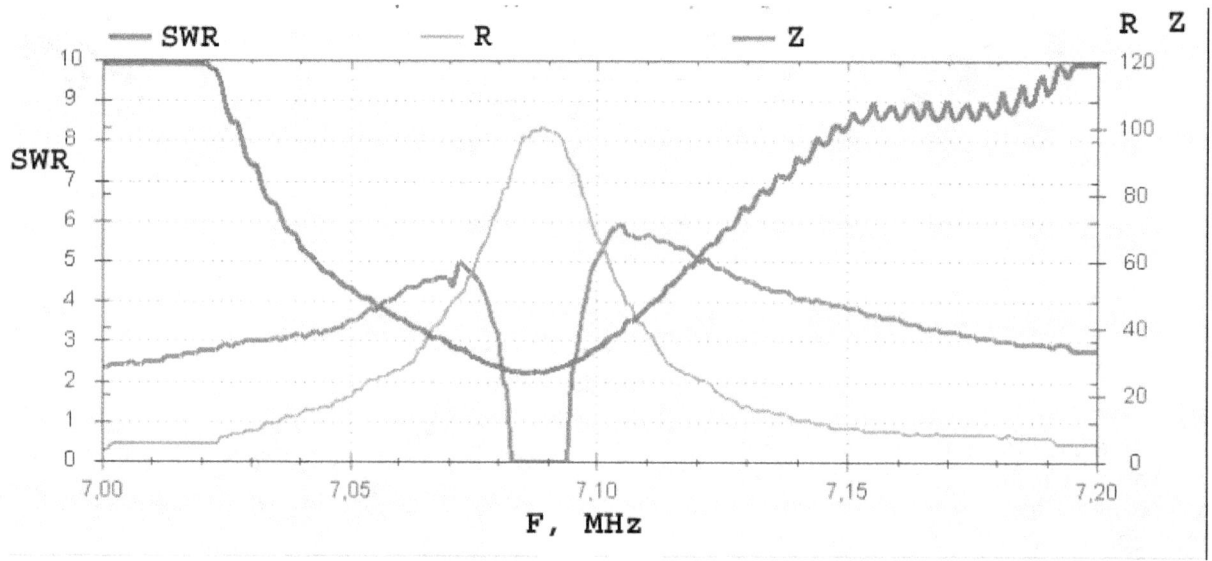

Picture 7 Parameters of the Field Antenna UA6AGW V.40.21

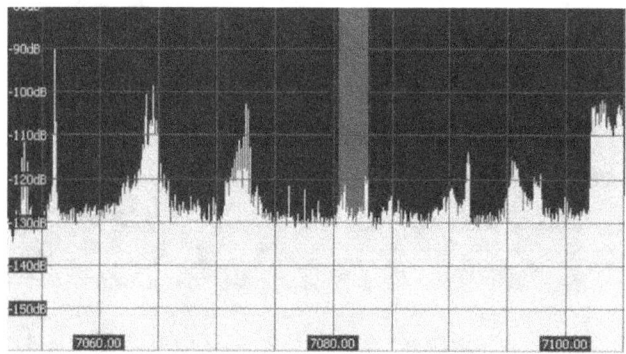

Picture 8 Screen Shot SDR Transceiver with Antenna UA6AGW V.40.02

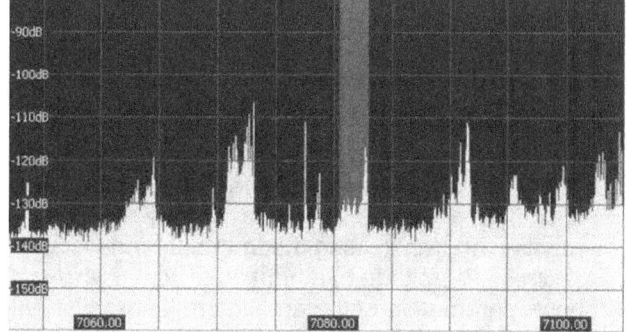

Picture 9 Screen Shot SDR Transceiver with Antenna UA6AGW V.40.21

Noise difference between the two antennas was from 5 up to 10- dB. Noise increased when Antenna UA6AGW V.40.21 was rotated in direction to the central part of the city. Then antenna UA6AGW V.40.21 was oriented to minimum of the noise. At the position the Antenna UA6AGW V.40.21 (compare to Antenna UA6AGW V.40.02) gives advantage at reception of weak signals.

Test at 4-meters height.

At height 4- meters above the ground antenna UA6AGW V.40.21 worked in the same way as at 6- meter height above the ground. No detuning in the resonance frequency was found. Antenna provided reception with low noise at day time period.

It was found very interesting property of the antenna UA6AGW V.40.21. At evening time the antenna provided reception of the nearest stations with lower level compare with Antenna UA6AGW V.40.02.

Stations, placed at radius 400- 500- km, were received lower then 10- dB. Stations, placed at radius 400- 800- km, were received lower then 5- dB. Stations placed at distance 1000- km and more from the antenna was received with the same level as with Antenna UA6AGW V.40.02.

Summary

Antenna UA6AGW V.40.21 could work at a small height. The antenna takes small room. Antenna made from low-cost materials. Antenna is easy to tuning and installations.

Reference
1. http://www.antentop.org/017/ua6agw_md_017.htm

73! UA6AGW

Shortened Antenna G5RV for 14- 50- MHz Bands

Alex Karakakan, UY5ON, Kharkov Ukraine

Shortened antenna G5RV is a variant classical G5RV with shortened radiation parts and shortened matching two-wire line. Amateur's Bands within 14- 50- MHz spread are mostly welcome for DX operation. So in most cases the antenna could provide good operation with DX-stations. Antenna takes small room.

Any limited space- balcony, windows are suitable for the antenna. Antenna may be placed in horizontal placement or similar to an I.V. Antenna could work at the 70- MHz band when proper transformer 4:1 is used. At some cases an ATU installed between coaxial cable and transceiver would be useful.

Credit Line: forum at cqham.ru

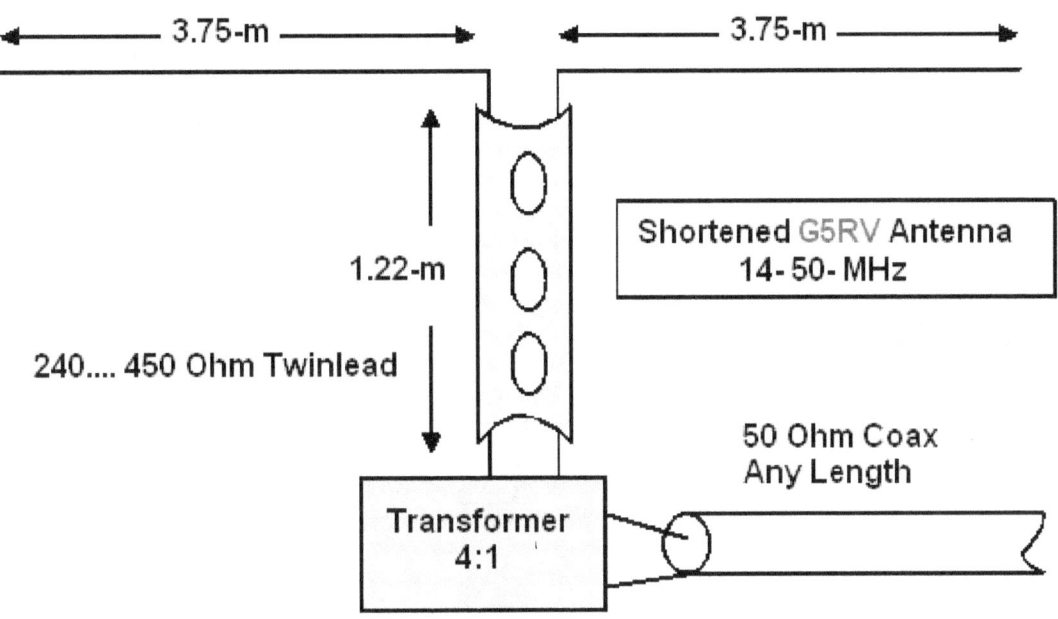

Shortened Antenna G5RV for 14- 50- MHz Bands

www.cqham.ru

Shortened Dipole Balcony Antenna for the 20-meter Band

Viktor Kovalensky, RN9AAA

Some years ago, just for fun, I made the Shortened Dipole Antenna for the 20- meter Band at my balcony. The antenna still exists and I use to it for my operation in the Air.

Figure 1 shows schematic of the antenna. As it is seen from the figure the antenna is shortened dipole antenna with lengthening inductor in each wire. Antenna fed by 50-Ohm coaxial cable. RF- Choke plus balun is installed at the cable of the antenna.

Picture 1 shows the antenna at installation on my balcony. RF-Choke installed on the coaxial cable of the antenna is very simple.

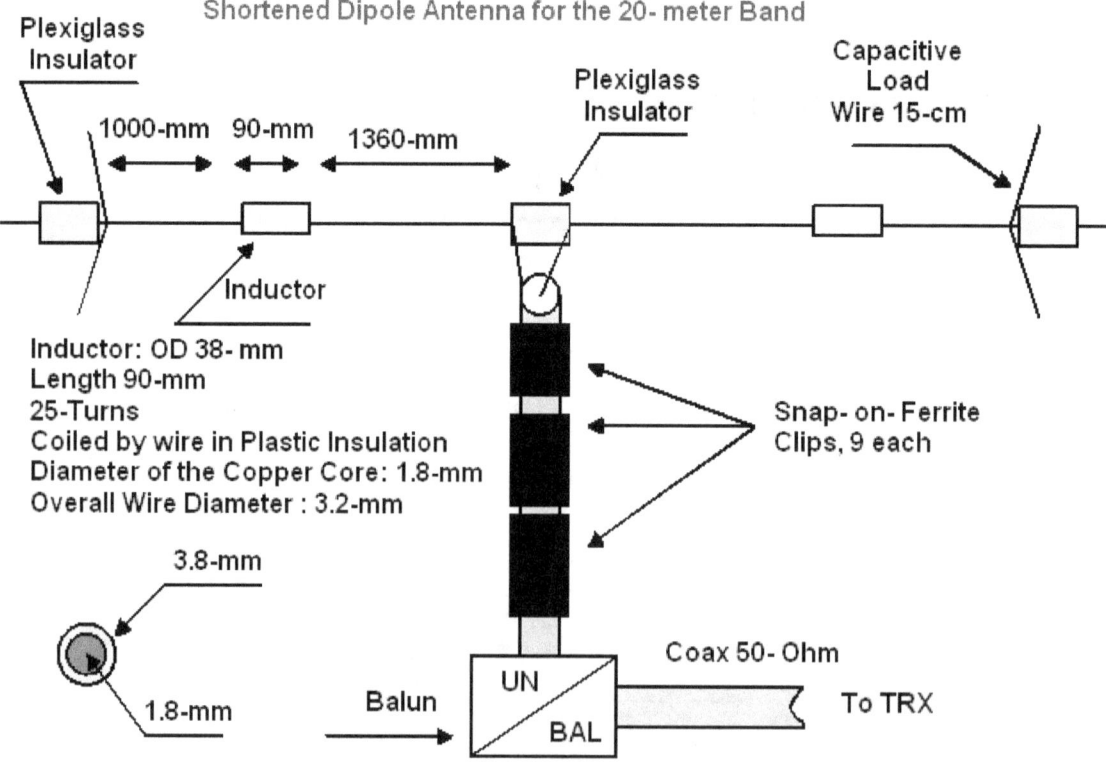

Figure 1 Shortened Dipole Antenna for the 20- meter Band

It is nine snap on ferrite clips that are snapped on to the coaxial cable. Quantity of the ferrite clips depends on whether the position of the coaxial cable is not interacted on to tuning of the antenna. When the needed quantity of the ferrite clips is found, install a balun. It is possible to find lots description "how to make a balun" in the internet. My balun also is very simple. It is four coils turned around a ferrite ring. It is possible use almost any ferrite ring with suitable sizes and permeability 100- 2000. **Picture 2** shows antenna analyzer MFJ- 259B connected to the tuned into resonance antenna. Antenna input impedance is 53- Ohm at 14.075- MHz.

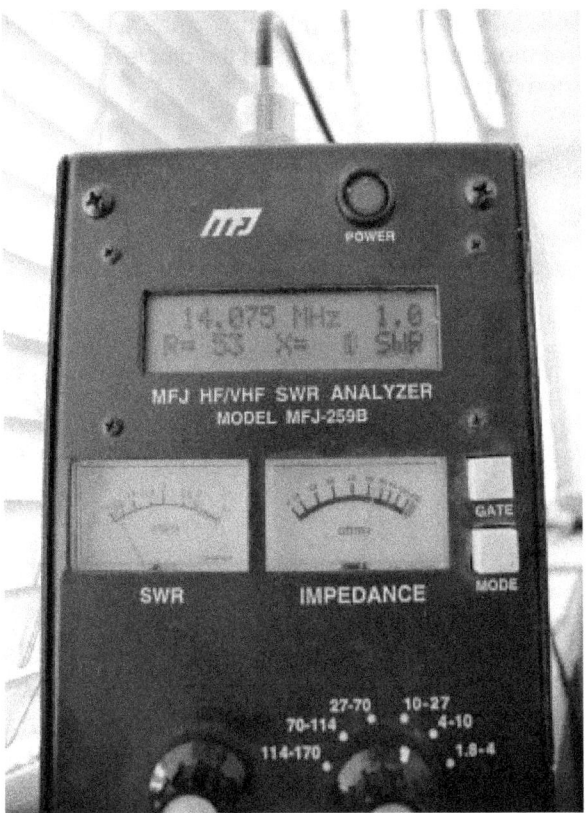

Picture 2 Antenna Analyzer MFJ- 259B Connected to the Tuned into Resonance Antenna

Design of the antenna is very simple. For the antenna it was used wire in diameter 0.7-mm (22- AWG). It is possible to use any suitable wire in diameter 0.5- 2.0- mm (25- 14- AWG). Homebrew insulators (made from plexiglass) used at the antenna. Lengthening inductor is wound around plastic tube in 38- mm (1-1/2") diameter. For the inductor it was used electro- technical wire in plastic insulation. Diameter of the copper core is 1.8-mm (13- AWG). Overall diameter of the wire in plastic insulation is 3.8- mm (7- AWG).

Picture 1 Shortened Dipole Antenna installed at my balcony

www.cqham.ru

Picture 3 shows the inductor. Capacitive load is installed at the ends of each wings of the dipole. **Picture 4** shows the capacitive load.

Antenna is tuned in to resonance by shortened of the capacitive loads. It needs cut the wire symmetrically at the both sides of the antenna. Install initially capacitive load in 20- cm- length.

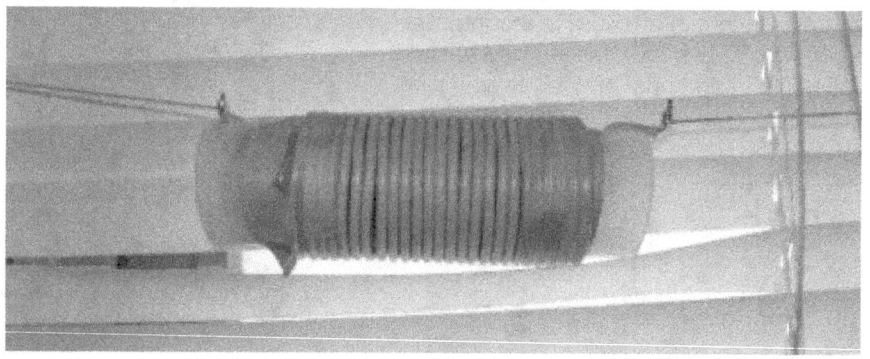

Picture 3 Inductor of the Shortened Dipole Antenna for the 20- meter Band

Picture 4 Capacitive Load at the Shortened Dipole Antenna for the 20- meter Band

Of course the antenna cannot be a "super- efficiency" antenna. However the antenna provides not bad operation in the Air by CW and JT65. **Picture 5** shows screen shot of monitor for JT65 for one of the days. **Reference 1** gives information how to make a shortened dipole antenna.

References
1. http://www.k7mem.com/Electronic_Note.../shortant.html

Picture 5 Screen Shot of Monitor for JT65

73! RN9AAA

FREE e- magazine edited by hams for hams
Devoted to Antennas and Amateur Radio
www.antentop.org

Simple Window Loop Antenna

Aleksandr Sterlikov, RA9SUS

The antenna was installed across wooden window frame. Perimeter of the loop in my case was 4.7- meter. Antenna could be tuned from 14 to 30- MHz. I fed the antenna 50- Ohm coaxial cable. However, also I used to 75- Ohm coaxial cable with success. **Figure 1** shows the antenna. At antenna feeding terminal a simple home- brew symmetrical device (several turns coaxial cable coiled on to big ferrite ring) was installed.

Loop was made from wire in 0.3- mm (28- AWG) diameter. Wire was attached to the window frame by scotch. Variable capacitor should be with big gap between plates. However it is depends on power. I used usual capacitor 12- 495- pF from old tube receiver when I run 30- Wtts at 14- MHz and 10- Wtts at 21 and 28- MHz. My antenna was installed at 4th- floor at 5- store concrete building. Window was at North- East side. While a short time at 20- meter band there were made QSOs with 36- countries. There were ex-USSR, USA, Canada, Japan, Europe, Asia. I worked by CW (50%), SSB (40%), and then RTTY and SSTV.

RA9SUS

The window loop antenna may be installed practically at any window frame. Frequency range the loop antenna depends on perimeter of the loop. Small loop would be tuned to high bands loop with long perimeter would be work at lower bands.

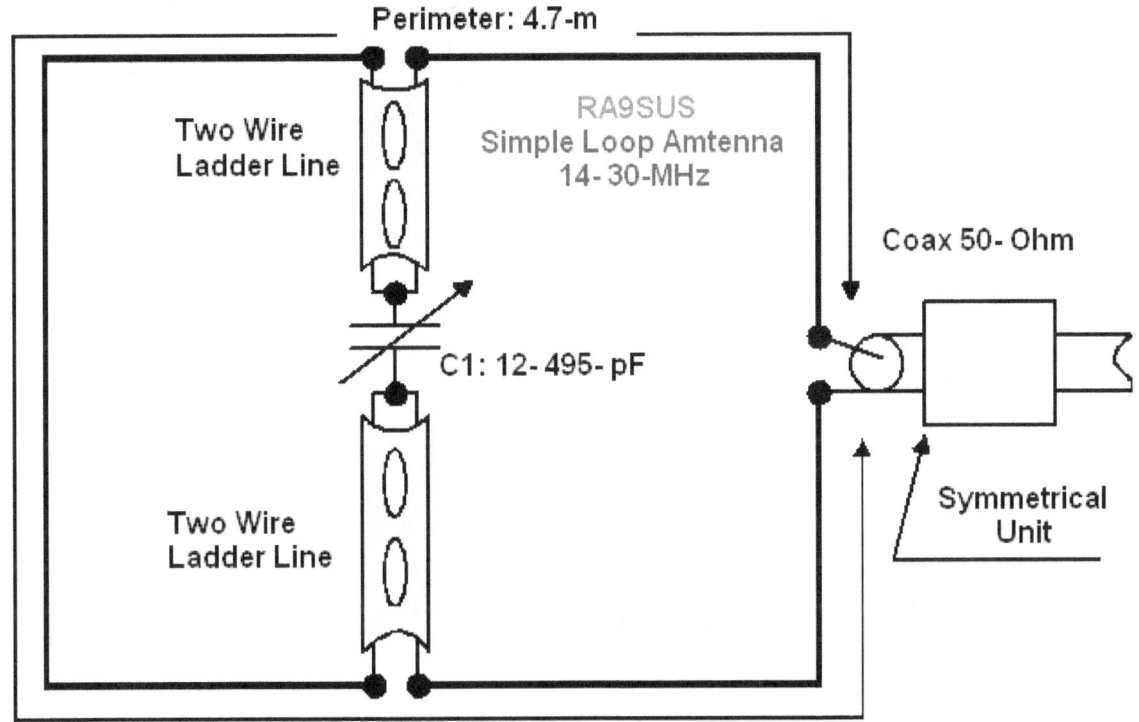

Figure 1 Simple Window Loop Antenna

Simple Folded Dipole Antenna for the 20- meter Band

Vladimir E. Tokarev, UA4HAZ

The antenna was designed for limited space. It may be installed on balcony, on fence or at backyard and masked on to rope for drying clothes. Antenna has input impedance cloth to 50- Ohm. **Figure 1** shows drawing of the antenna.

Folded sides of the antenna may be placed vertically, horizontally or by another way. Input impedance and resonance frequency of the antenna depends on to position of the antenna on to ground and on to different subjects. So antenna should be tuned in the place of its installation. You may compare parameters of the antenna installed on height of 3-meters abve the ground (**Figure 2**, **Figure 3**, **Figure 4**) with parameters of the antenna in free space (**Figure 5**, **Figure 6**, **Figure 7**).

MMANA file for the antenna may be downloaded at:
http://www.antentop.org/018/ua4haz_018.htm

73! UA4HAZ

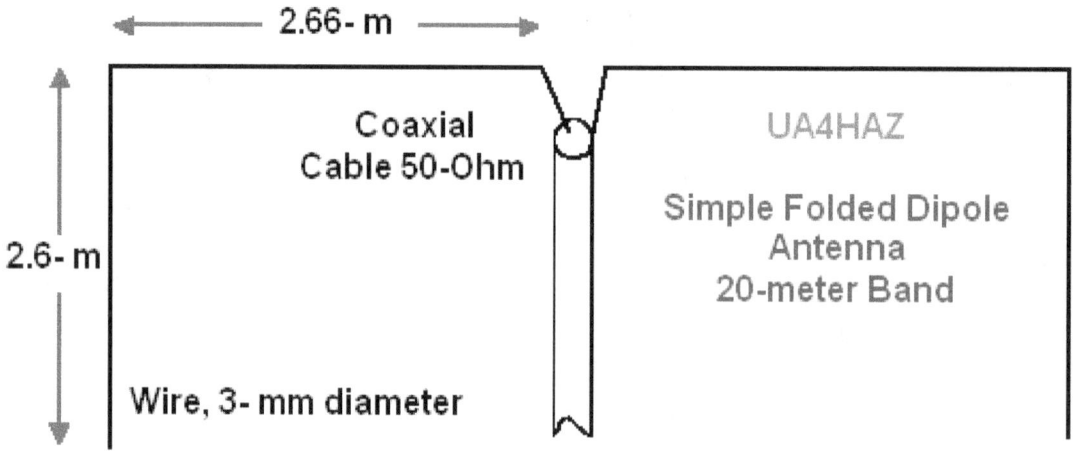

Figure 1 Simple Folded Dipole Antenna for the 20- meter Band

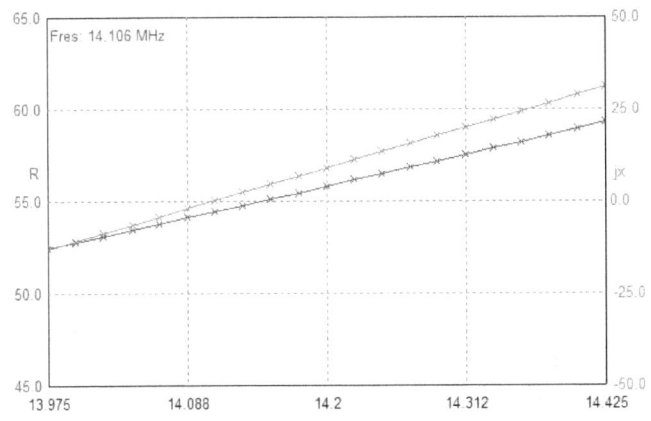

Figure 2 Input impedance of the Simple Folded Dipole Antenna installed at 3- meters above the ground

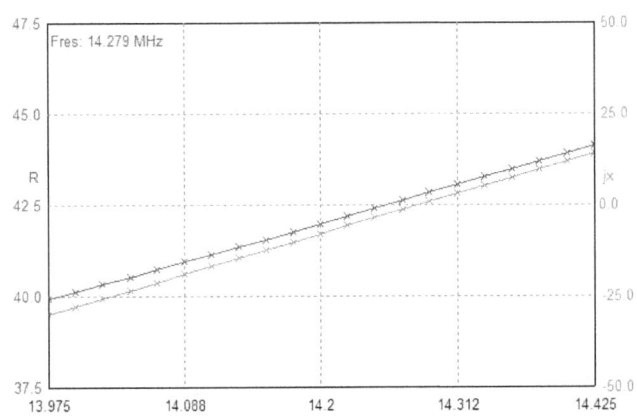

Figure 5 Input impedance of the Simple Folded Dipole Antenna in free space

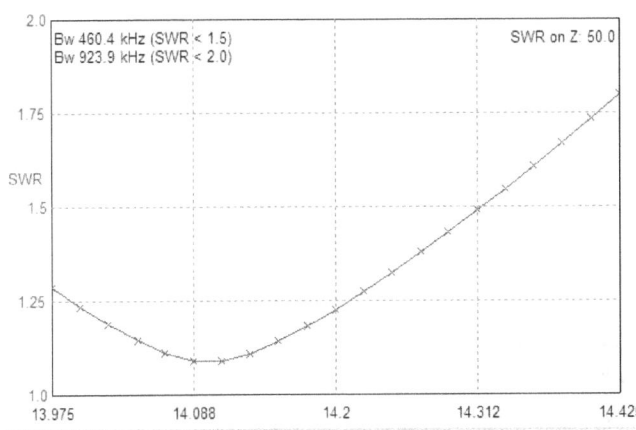

Figure 3 SWR of the Simple Folded Dipole Antenna installed at 3- meters above the ground

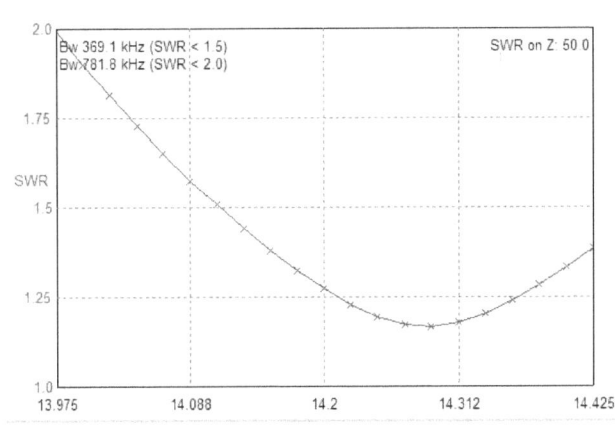

Figure 6 SWR of the Simple Folded Dipole Antenna in free space

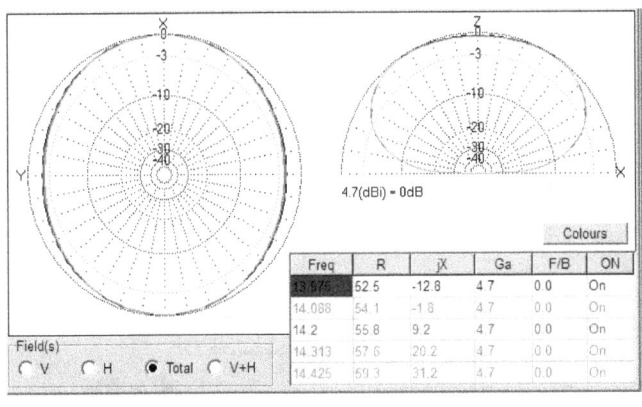

Figure 4 DD of the Simple Folded Dipole Antenna installed at 3- meters above the ground

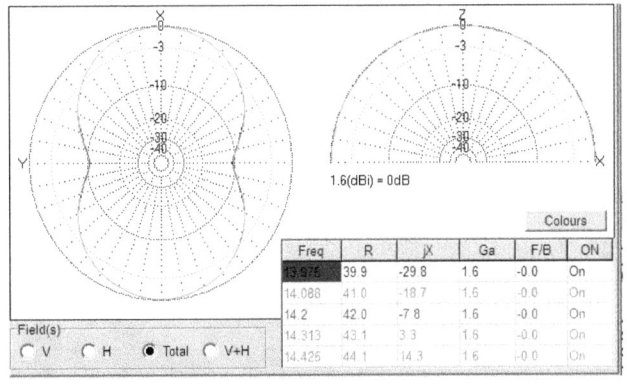

Figure 7 DD of the Simple Folded Dipole Antenna in free space

Simple Wire Antenna for All HF- Bands

Vladimir Fursenko, UA6CA

The simple wire antenna works well from 160- to 10 meter. The antenna may be tuned (to needed amateur band) at a shack. Antenna contains only one tuning parts- it is a variable capacitor 10- 200- pF. An inductor (near 3... 5- micro- Henry) is switched in serial with the antenna at the 40- meter Band. **Figure 1** shows schematic diagram of the Simple Wire Antenna.

Antenna has wide pass- band that covered all amateur band that is used by the antenna. Good grounding is needed for the antenna. The Simple Wire Antenna was used at station UA6CA with a 4.5- Watt QRP- Transceiver. Antenna shows good result at the operation.

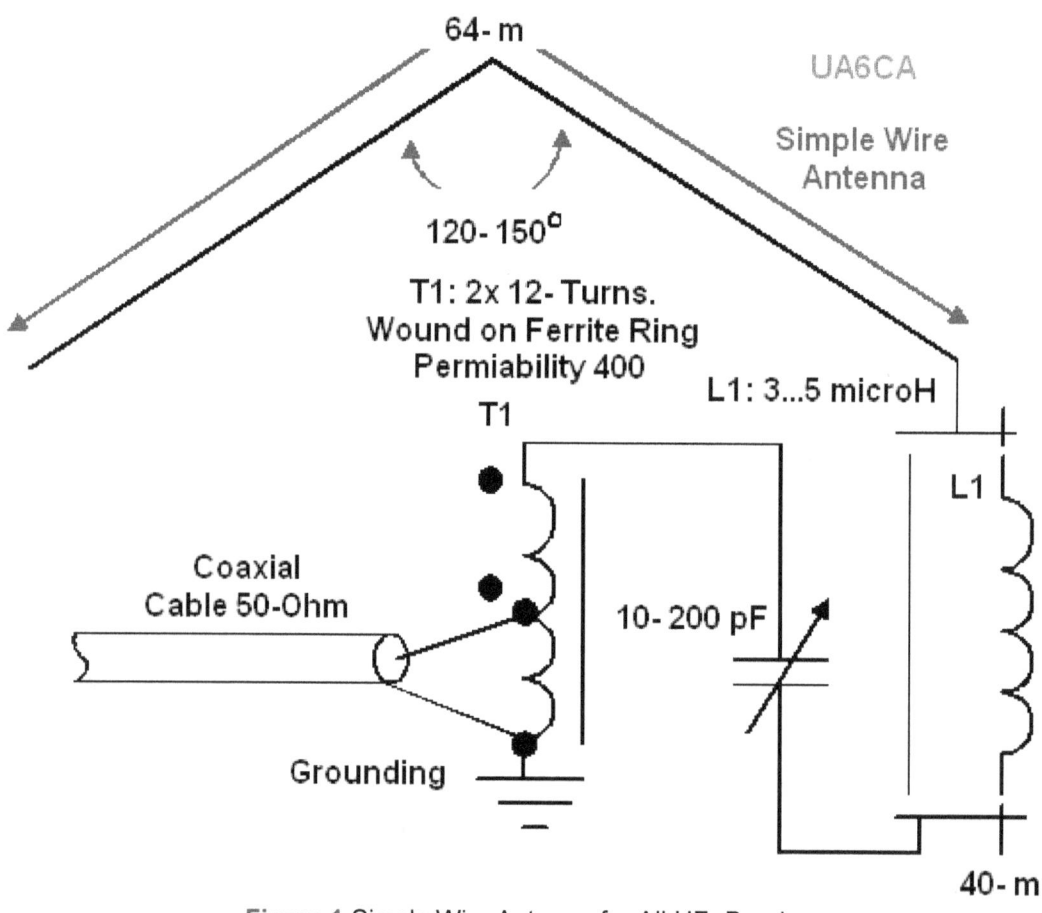

Figure 1 Simple Wire Antenna for All HF- Bands

FREE e- magazine edited by hams for hams
Devoted to Antennas and Amateur Radio
www.antentop.org

Twin Triangle Antenna for 10-meter Band

Yuriy Kondrat'ev, UA1ZAS

Credit Line: Radio # 2, 1977, p.: 19

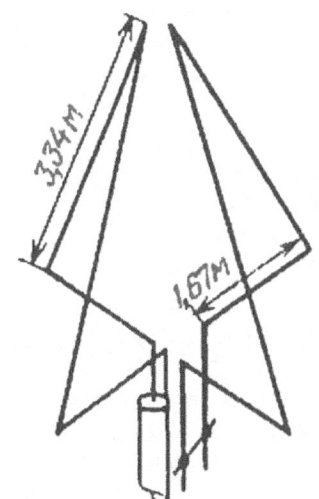

The antenna is used at the 10-meter Band. Antenna made from two wires triangles. The triangles are fastened to my Ground Plane antenna for the 20-meter band. The triangles do not influenced to the ground plane.

Tuning of the antenna is simple. SWR of the antenna tune at first by length of the radiated triangle. Then stub is tuned to minimum back radiation. The efficiency of the antenna is close to antenna G4ZU. Antenna may be fed by 50- or 75- Ohm Coaxial cable.

Compact Twin Delta Antenna for 80- and 40-meter Bands

John J. Schultz, W2EEY/7

Credit Line: Old Man # 4, 1976

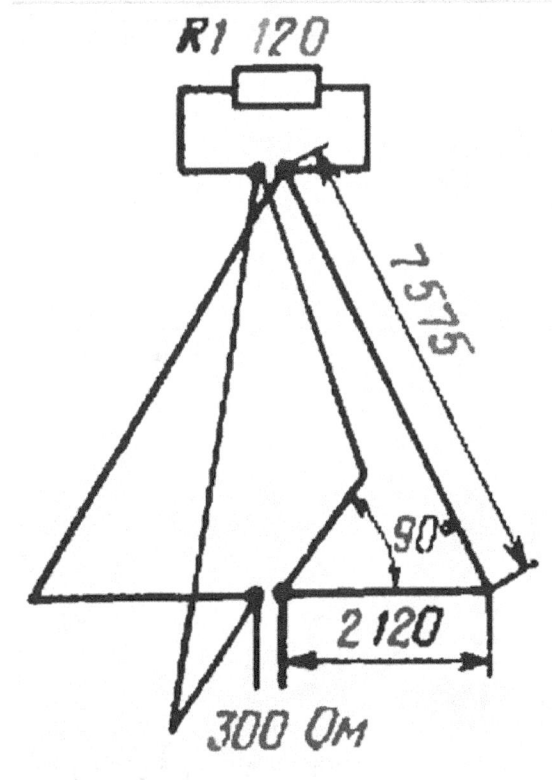

The antenna has wide broadband at the 80- and 40-meter Bands. So the antenna does not require any tuning. Antenna is simple in design and takes lots room for installation. Antenna radiates a vertical polarisation wave with almost circle DD in horizontal plane. Antenna is not critical to the sizes. It could be got good result at perimeter of each triangle lambda/4 at lower band at the antenna. However, the antenna is still working when perimeter of the each triangle is lambda/8 at the lower band of the antenna.

Antenna is fed by 300- Ohm ladder line. However, for the antenna feeding it is possible use a coaxial cable if the top load resistor would be changed (resistance should be decreased).

Antenna was made practically and hanged up at height 1-meter above the ground. SWR of the antenna was less the 2.0: 1.0 at the bands. SWR may be decreased by changing resistance the top resistor (for local conditions).

Delta Antenna for 80-, 40-, 20- and 15- meter Bands

The publication is devoted to the memory UR0GT.

Credit Line: Forum from:
www.cqham.ru

By: Nikolay Kudryavchenko, UR0GT

The Delta Antenna has perimeter 86- meter. Antenna has resonances at four amateur bands. There are 80, 40, 20 and 15 meter. However, input impedance at the bands not allows use a 50- Ohm coaxial cable to feed the antenna with low SWR at all the bands. Best solution is to use a 100- Ohm coaxial cable. Then simple ATU to match the antenna system with transceiver.

By the way, common use TV 75- Ohm coaxial cable may be used to feed the antenna. Two wires open line allows use the antenna at all amateur HF bands. At this case an ATU should be used to match the antenna system,

Figure 1 shows three version of the antenna. Each of the versions has some advantages and disadvantages.

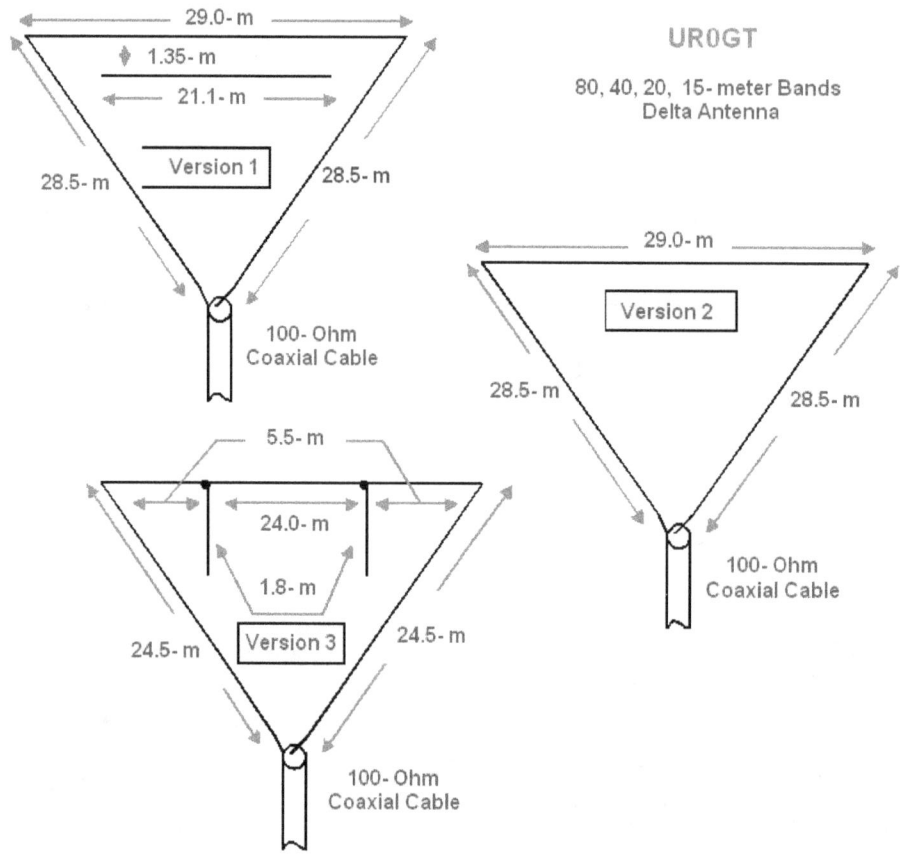

Figure 1 Design of the Delta Antenna for 80-, 40-, 20- and 15- meter Bands

Figure 2 shows SWR at 80- meter Band for version 1.
Figure 3 shows SWR at 40- meter Band for version 1.
Figure 4 shows SWR at 20- meter Band for version 1.
Figure 5 shows SWR at 15- meter Band for version 1.
Antenna is placed at height 15- meter above the ground.

Figure 6 shows SWR at 80- meter Band for version 2.
Figure 7 shows SWR at 40- meter Band for version 2.
Figure 8 shows SWR at 20- meter Band for version 2.
Figure 9 shows SWR at 15- meter Band for version 2.
Antenna is placed at height 15- meter above the ground.

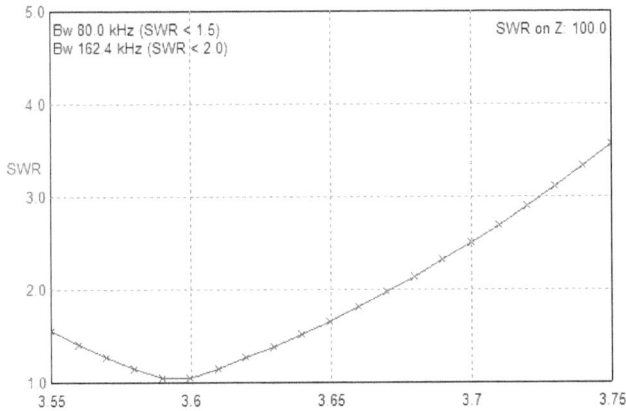

Figure 2 SWR at 80- meter Band for version 1

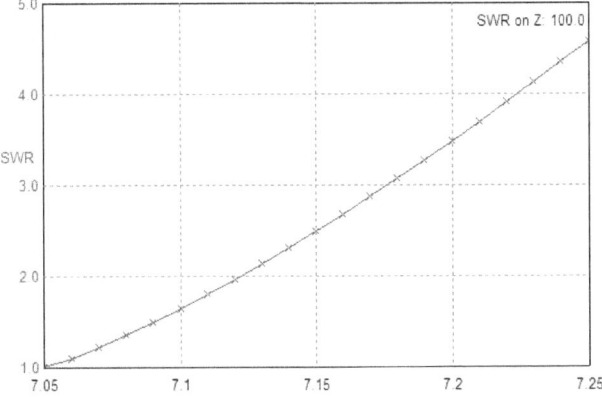

Figure 3 SWR at 40- meter Band for version 1

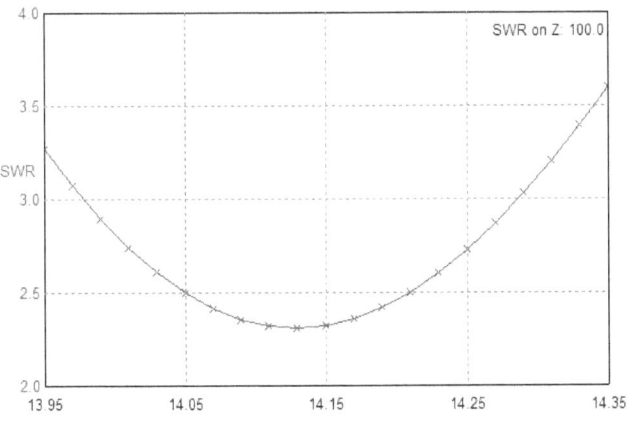

Figure 4 SWR at 20- meter Band for version 1

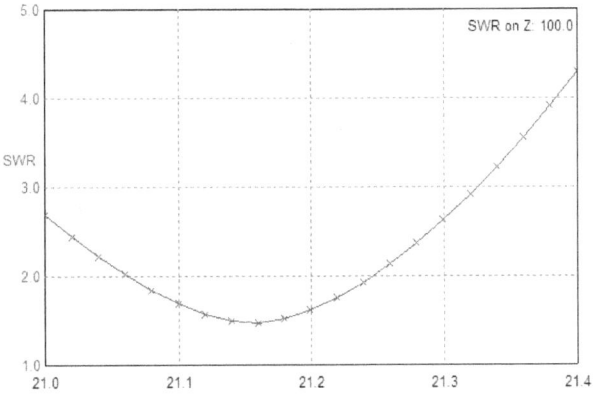

Figure 5 SWR at 15- meter Band for version 1

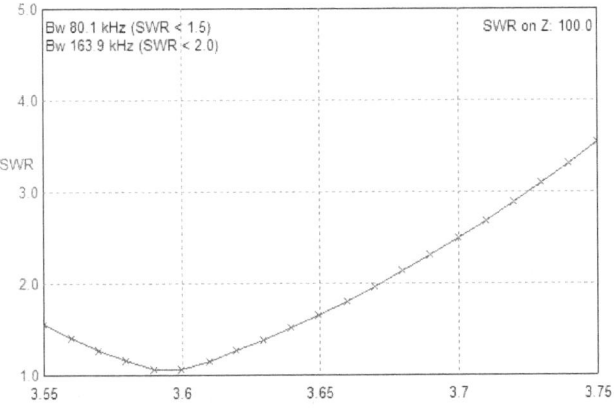

Figure 6 SWR at 80- meter Band for version 2

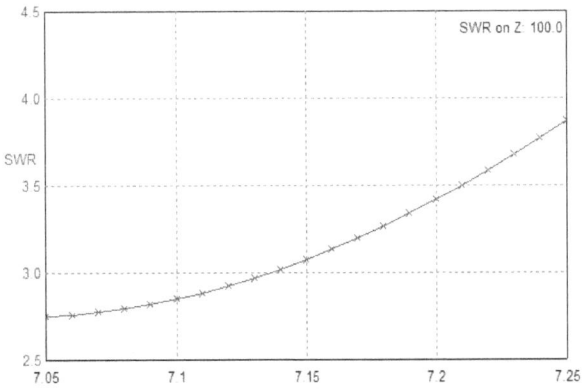

Figure 7 SWR at 40- meter Band for version 2

Figure 10 shows SWR at 80- meter Band for version 3.
Figure 11 shows SWR at 40- meter Band for version 3.
Figure 12 shows SWR at 20- meter Band for version 3.
Figure 13 shows SWR at 15- meter Band for version 3.
Antenna is placed at leaning position at the ground.

The MMANA model of the Delta Antenna for 80-, 40-, 20- and 15- meter Bands may be loaded: http://www.antentop.org/018/ur0gt_delta_018.htm

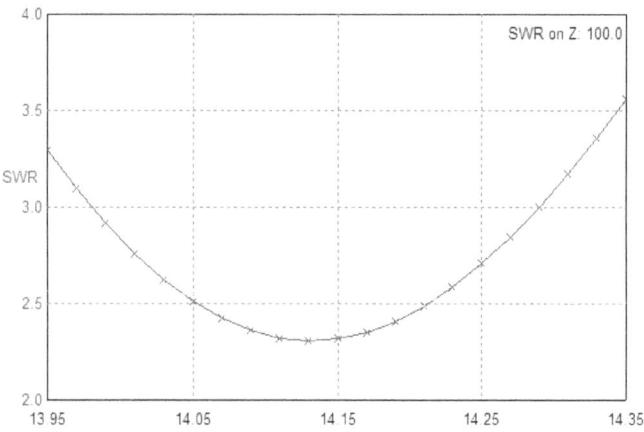

Figure 8 SWR at 20- meter Band for version 2

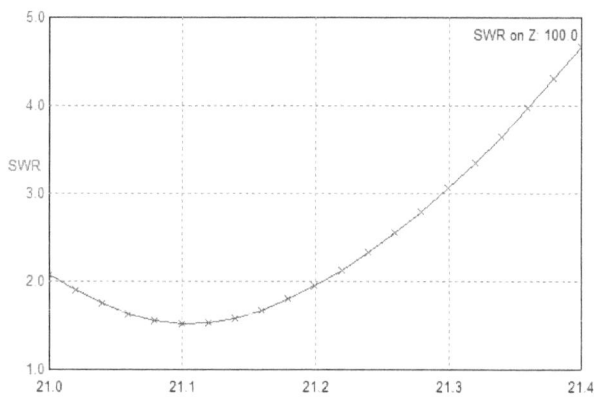

Figure 9 SWR at 15- meter Band for version 2

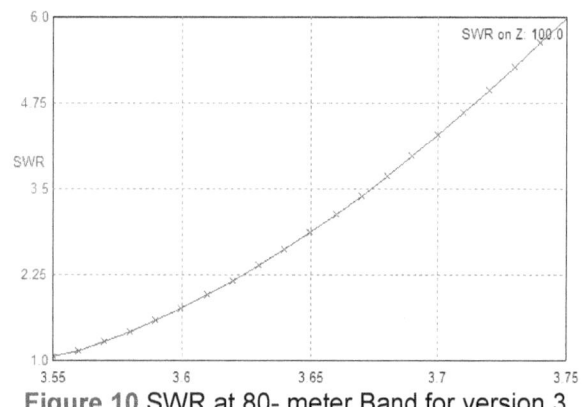

Figure 10 SWR at 80- meter Band for version 3

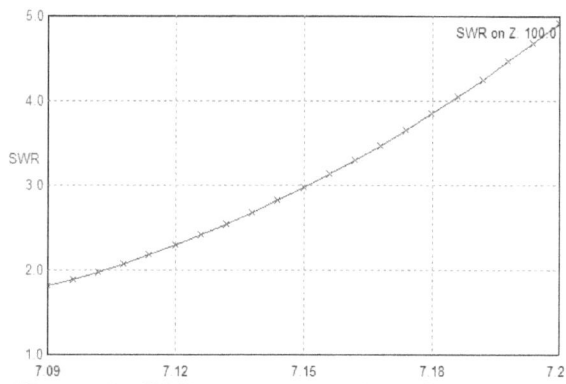

Figure 11 SWR at 40- meter Band for version 3

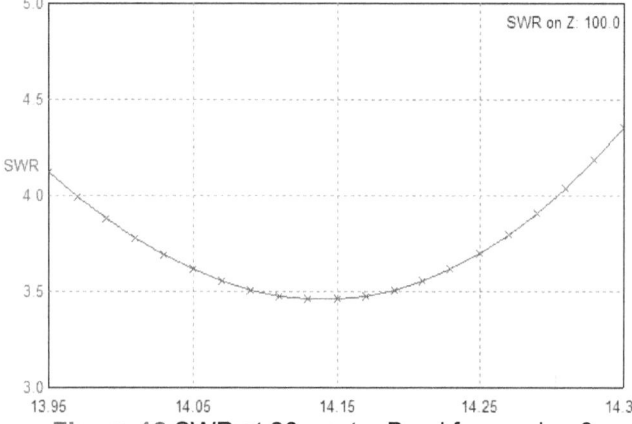

Figure 12 SWR at 20- meter Band for version 3

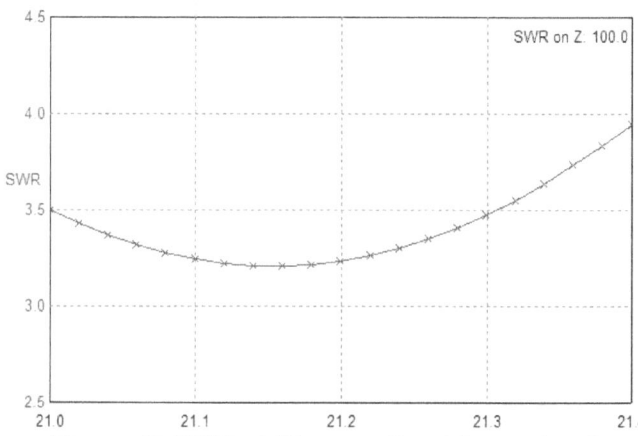

Figure 13 SWR at 15- meter Band for version 3

73 Nick

Windom UA6CA for 80-, 40-, 20- and 10- meter Bands

Vladimir Fursenko, UA6CA

Antenna was installed at the edge of the roof of the 5-store house. Transmitter was placed at the first floor. For improving of the efficiency of the antenna a grounding "mirror" wire was dug in the ground. Mirror wire was in plastic insulation. Ends of the mirror wire and connection to the wire were insulated from the ground. **Figure 1** shows the antenna.

Antenna was used with transmitter with Pi- Filter at transmitting output. The Pi- Filter could match well the antenna system. Antenna worked good at the 80, 40, 20 and 10- meter Bands. The antenna was used on UA6CA station at 1970- 1972 years.

73! UA6CA

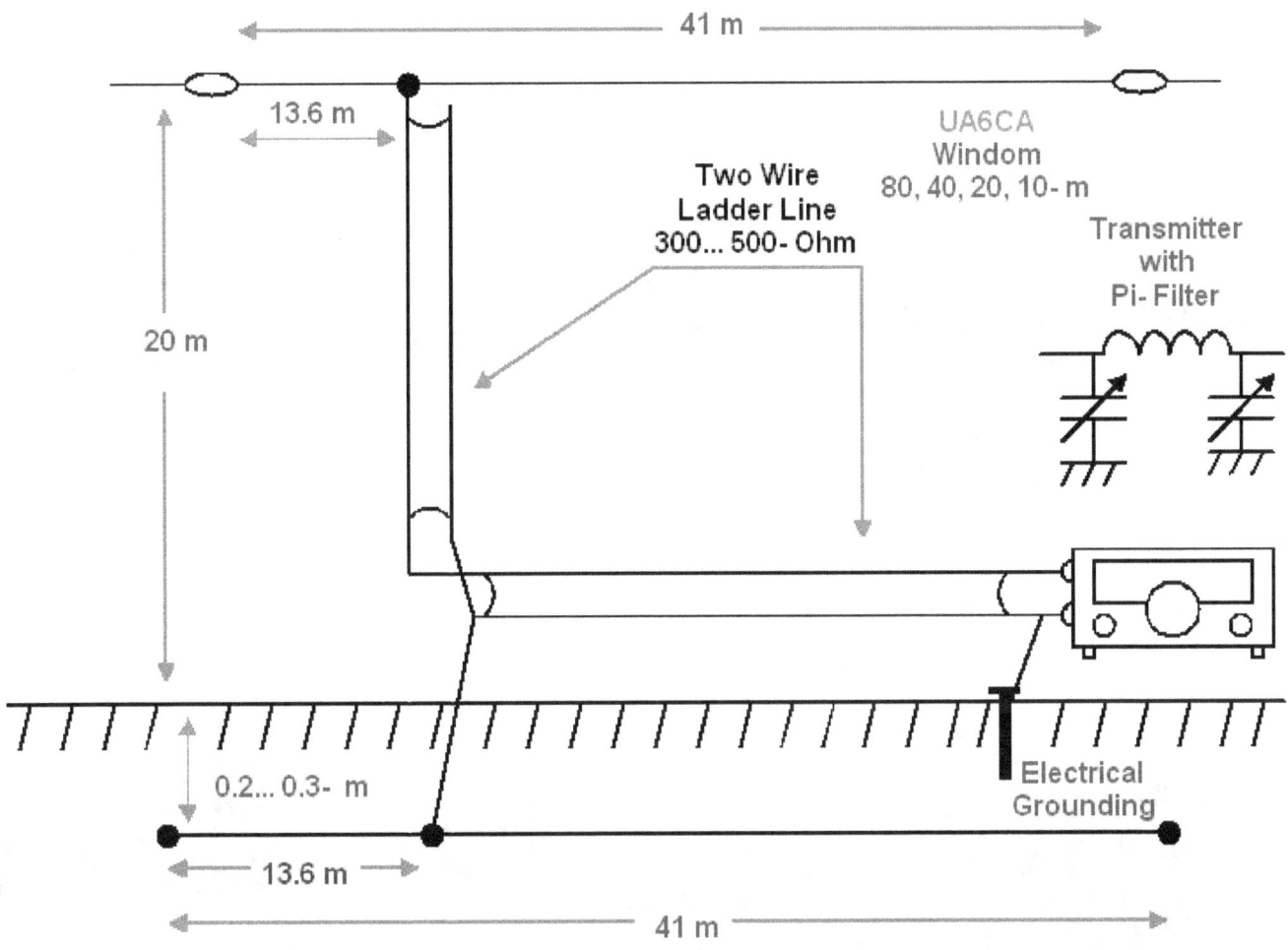

Figure 1 Windom UA6CA for 80-, 40-, 20- and 10- meter Bands

Air Plane HF Antennas

Boeing 747: Antenna at the Top of the Tail

Russian AN-12
Antenna at the Top of Cockpit

Ukrainian AN- 24: Antenna – Cockpit- Tail

Ukrainian AN- 26: Antenna – Cockpit- Tail

Russian Helicopter:
Antenna Cockpit- Tail

VHF- UHF Antennas

www.antentop.org

Car Antenna for the 435- MHz

The publication is devoted to the memory UR0GT.

Credit Line: Forum from:
www.cqham.ru

By: Nikolay Kudryavchenko, UR0GT

The Car Antenna for the 435- MHz has very good parameters- circle DD in Horizon Plane and small angle lobe at Vertical Plane. Antenna has input impedance 50- Ohm. Antenna has wide pass band.

Ground of the antenna should be connected to metal part of the Magnet Base. The capacitance between the magnet Base and Car body is enough for good performance of the antenna. Antenna should be placed at the center body of the car- it may be roof of trunk of the car.

Of course, the antenna may be done in base variant- with usual lambda/4 counterpoises. **Figure 1** shows two versions of the antenna- - short and long ones.

Figure 2 shows impedance of the short Car Antenna for the 435- MHz. **Figure 3** shows SWR of the short Car Antenna for the 435- MHz. **Figure 4** shows DD of the short Car Antenna for the 435- MHz.

Figure 5 shows impedance of the long Car Antenna for the 435- MHz. **Figure 6** shows SWR of the long Car Antenna for the 435- MHz. **Figure 7** shows DD of the long Car Antenna for the 435- MHz.

The MMANA file of the Car Antenna for the 435- MHz may be loaded at: http: //
www.antentop.org/018/car_antenna_018.htm

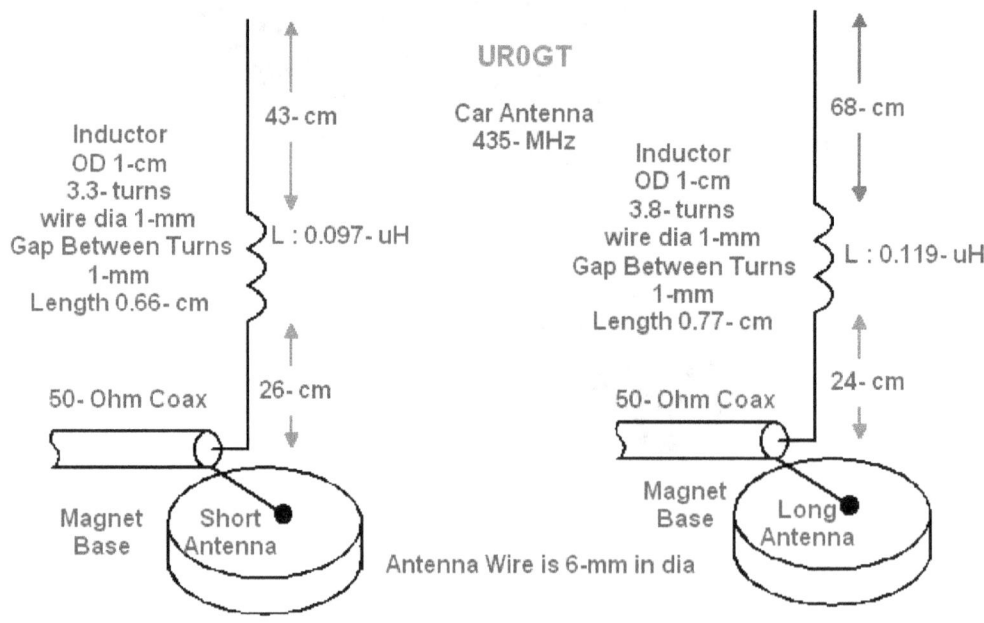

Figure 1 Car Antenna for the 435- MHz

www.cqham.ru

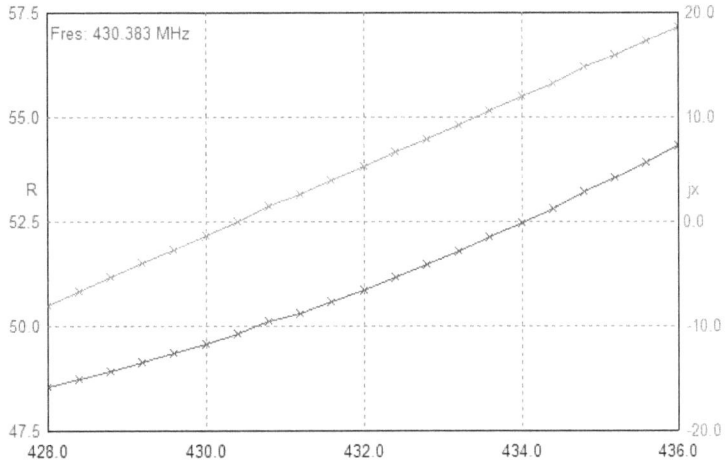

Figure 2 impedance of the Short Car Antenna for the 435- MHz

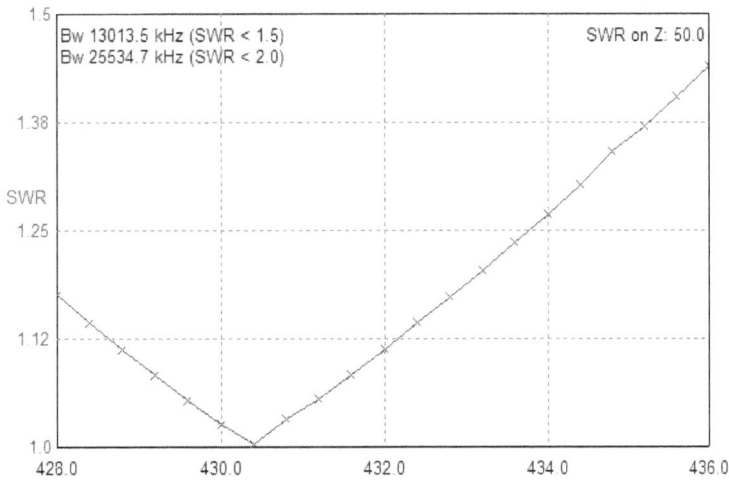

Figure 3 SWR of the Short Car Antenna for the 435- MHz

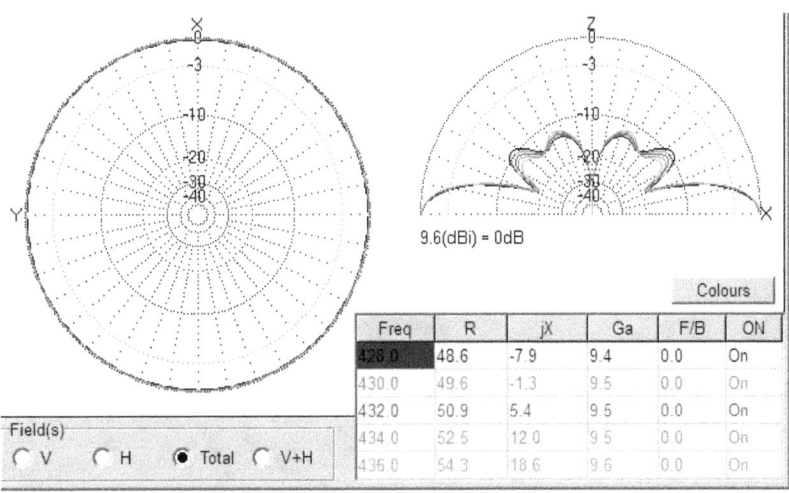

Figure 4 DD of the Short Car Antenna for the 435- MHz

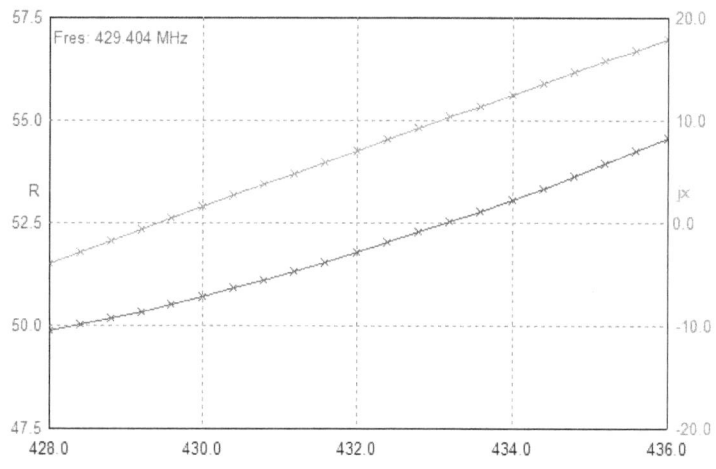

Figure 5 impedance of the Long Car Antenna for the 435- MHz

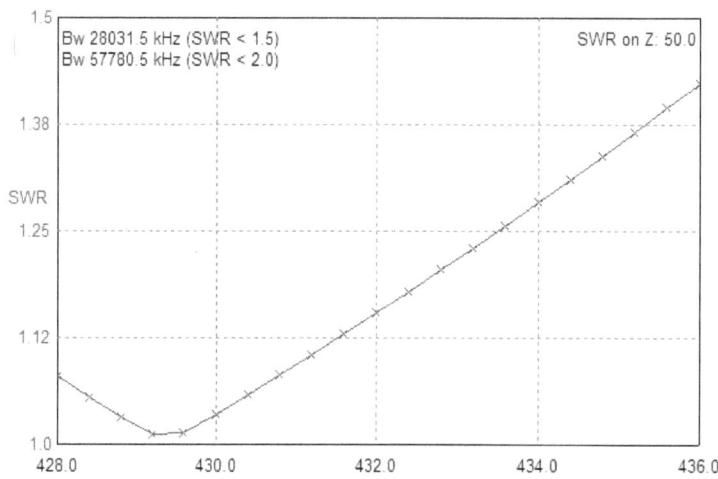

Figure 6 SWR of the Long Car Antenna for the 435- MHz

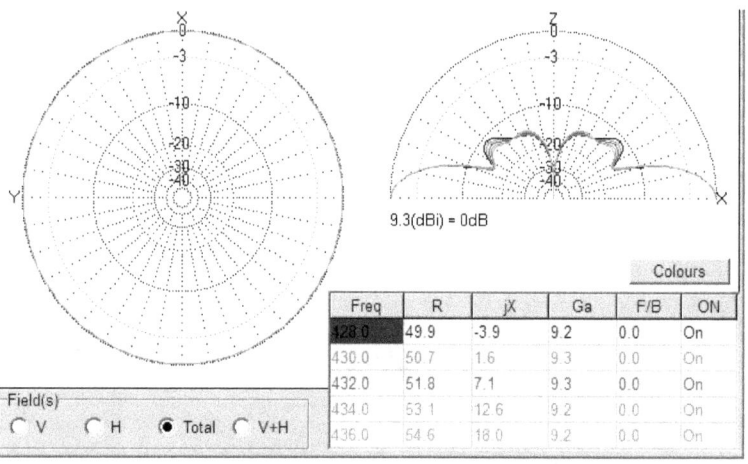

Figure 7 DD of the Long Car Antenna for the 435- MHz

73 Nick

Antenna for 2- meter Band, LPD (433), 70-cm Band and for RMR (446)

Credit Line: Forum from:
radioscanner.ru

By: Igor Vakhreev, RW4HFN

It is very simple antenna that allows works at several frequencies bands with low SWR. The antenna is enough broadband that does not required hold strictly sizes at the design. **Figure 1** shows design of the antenna. Made according to the **Figure 1** antenna does not require any tuning.

Figure 2 shows impedance of the antenna at the 2-meter Band. **Figure 3** shows SWR of the antenna at the 2- meter Band. **Figure 4** shows DD of the antenna at the 2- meter Band.

Figure 5 shows impedance of the antenna at the 70-cm Band. **Figure 6** shows SWR of the antenna at the 70-cm Band. **Figure 7** shows DD of the antenna at the 70-cm Band.

The MMANA file of the Antenna for 2-meter Band, LPD (433), 70- cm Band and for RMR (446) may be loaded at: http: // www.antentop.org/018/rw4hfn_018.htm

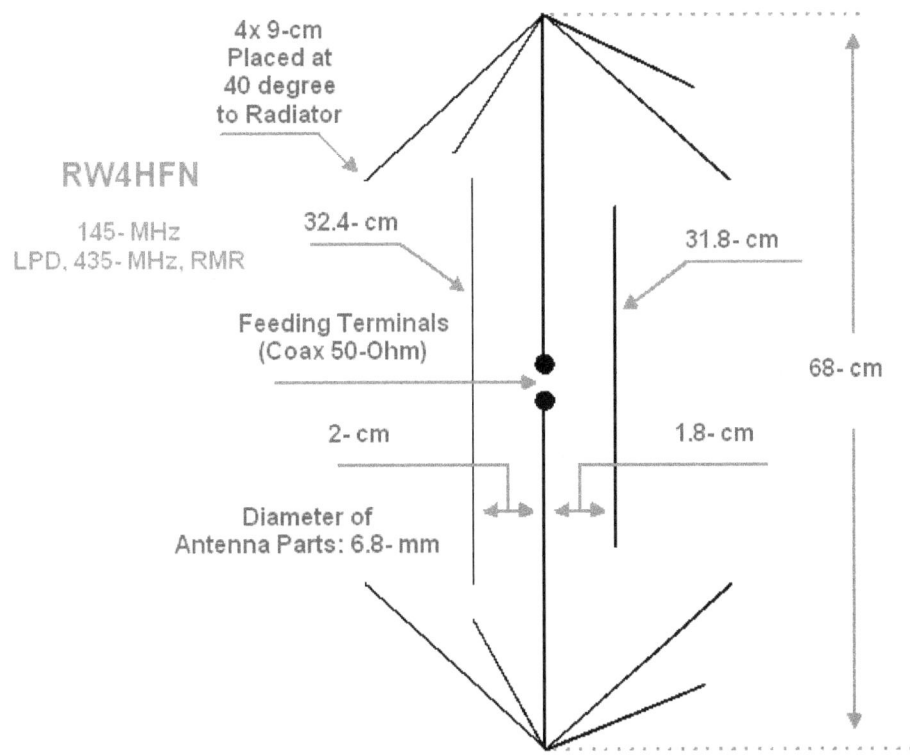

Figure 1 Antenna for 2-meter Band, LPD (433), 70- cm Band and for RMR (446)

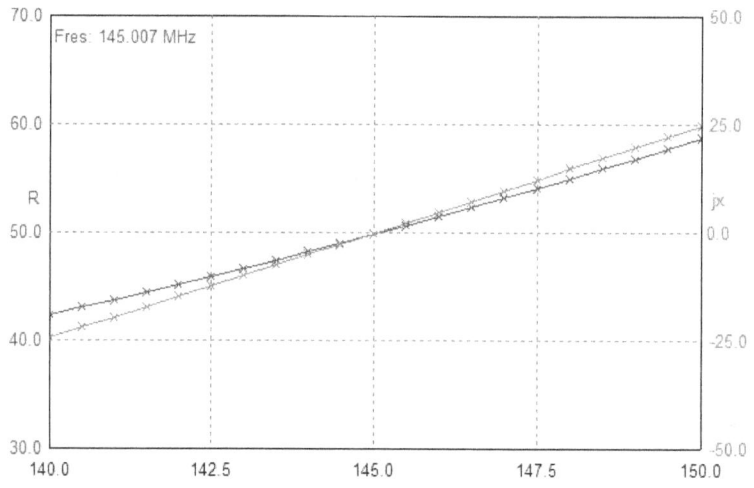

Figure 2 impedance of the antenna at the 2- meter Band

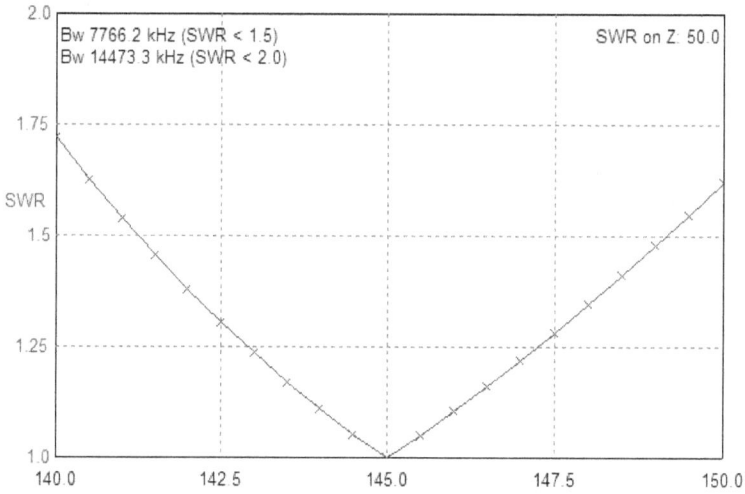

Figure 3 SWR of the antenna at the 2- meter Band

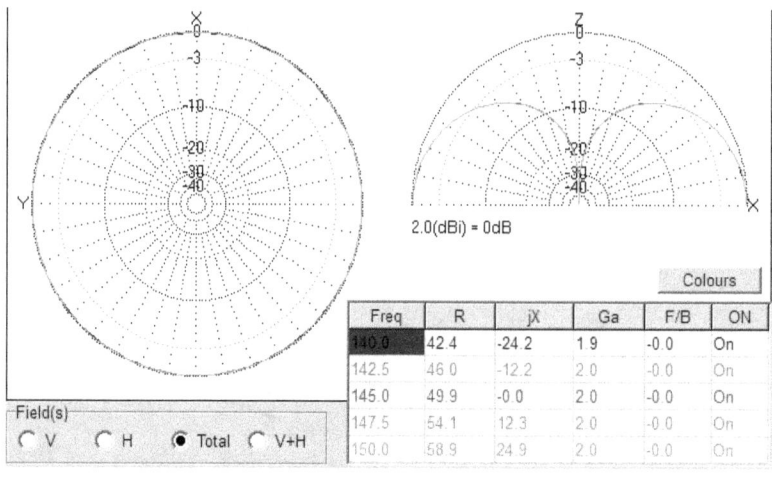

Figure 4 DD of the antenna at the 2- meter Band

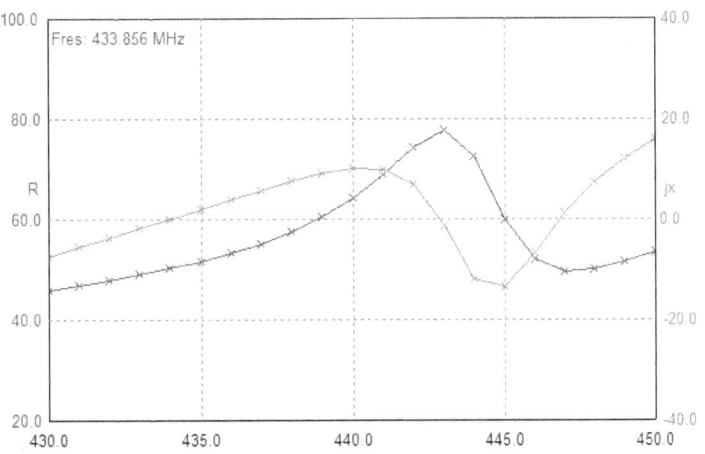

Figure 5 impedance of the Antenna at the 70-cm Band

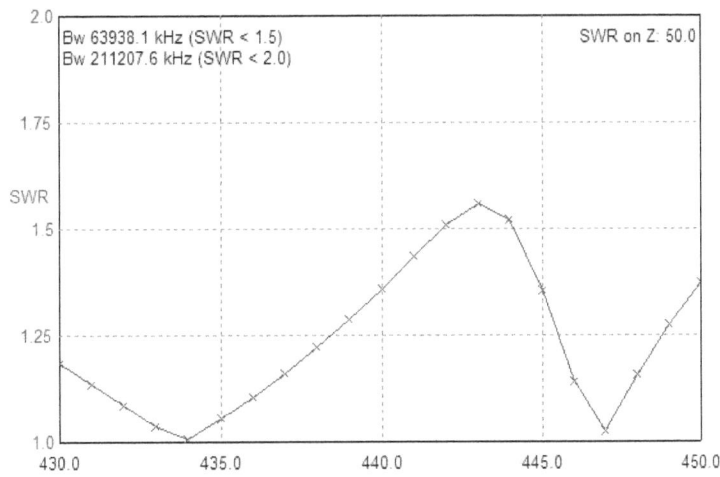

Figure 6 SWR of the Antenna at the 70-cm Band

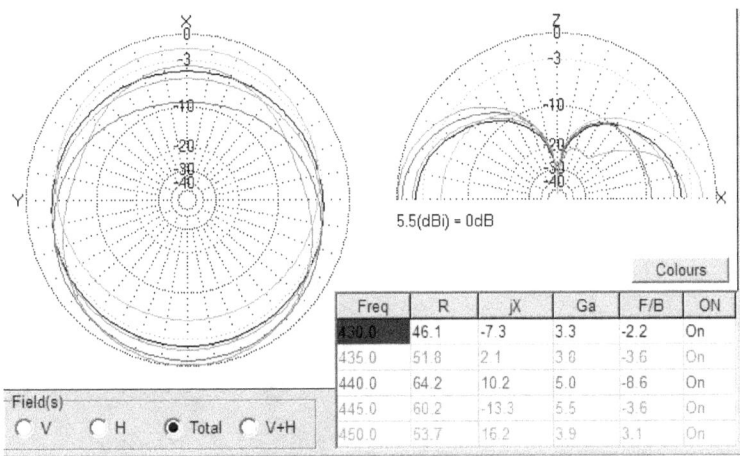

Figure 7 DD of the Antenna at the 70-cm Band

73! RW4HFN

Twin Delta Antenna for the 2- meter Band

The publication is devoted to the memory UR0GT.

Credit Line: Forum from:
www.cqham.ru

By: Nikolay Kudryavchenko, UR0GT

The simple Twin Delta Antenna works fine at the 2- meter Band. The antenna is enough broadband that does not required hold strictly sizes at the design. Antenna could be made from wide range diameters of wire – 2… 10- mm would be good. Antenna does not require any symmetrical devices.

The antenna has closed loop so it is low noise antenna and the antenna has some protection from static and lightning. Antenna fed through 50- Ohm coaxial cable.

Figure 1 shows design of the antenna. **Figure 2** shows impedance of the antenna (antenna placed at 7- meter above the real ground). **Figure 3** shows SWR of the antenna (antenna placed at 7- meter above the real ground). **Figure 4** shows DD of the antenna (antenna placed at 7- meter above the real ground).

The MMANA file of the Twin Delta Antenna may be loaded: http: // www.antentop.org/018/twin_delta_018.htm

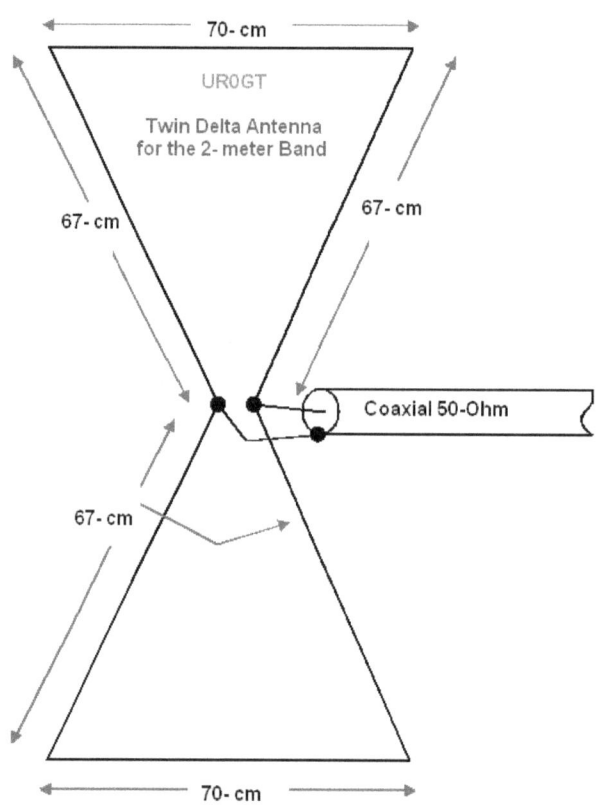

Figure 1 Twin Delta Antenna for the 2- meter Band

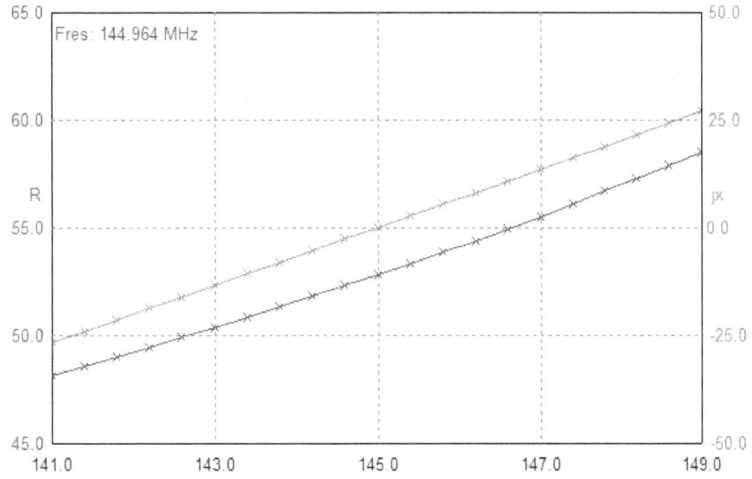

Figure 2 impedance of the Twin Delta Antenna for the 2- meter Band

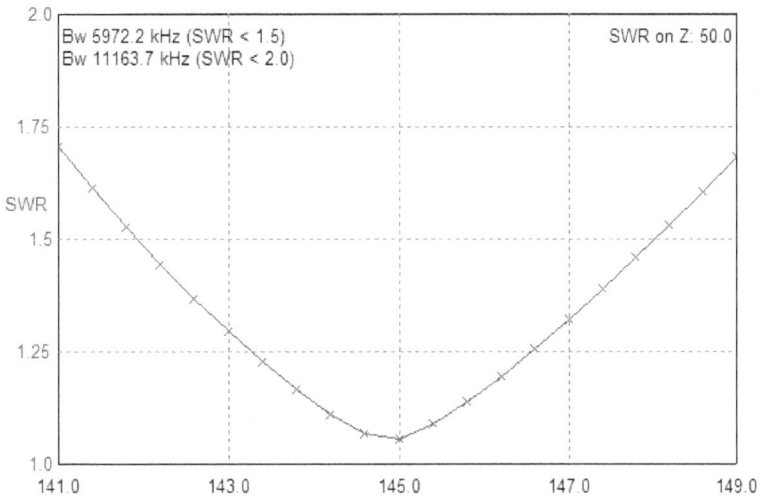

Figure 3 SWR of the Twin Delta Antenna for the 2- meter Band

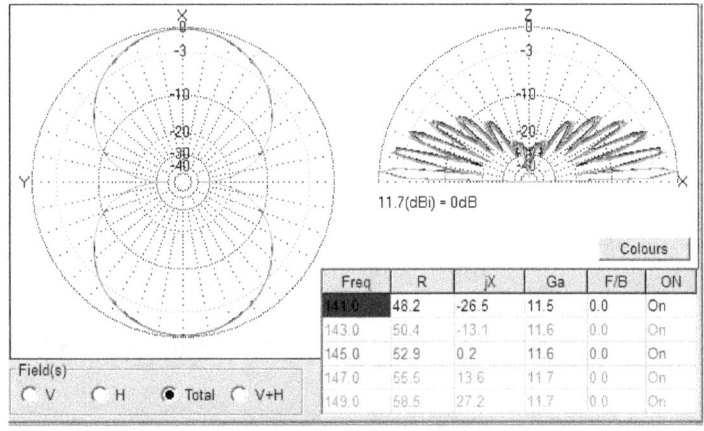

Figure 4 DD of the Twin Delta Antenna for the 2- meter Band

73 Nick

Conversion Auto CB Antenna HUSTLER-1C-100 to Antenna for the 2- meter Band

Igor Mishin, UT3IM

It is very easy convert Auto CB-Antenna HUSTLER – 1C-100 (on magnet base) to antenna working at 2- meter Band. **Figure 1** shows the conversation.

Credit Line: Forum at www.cqham.ru

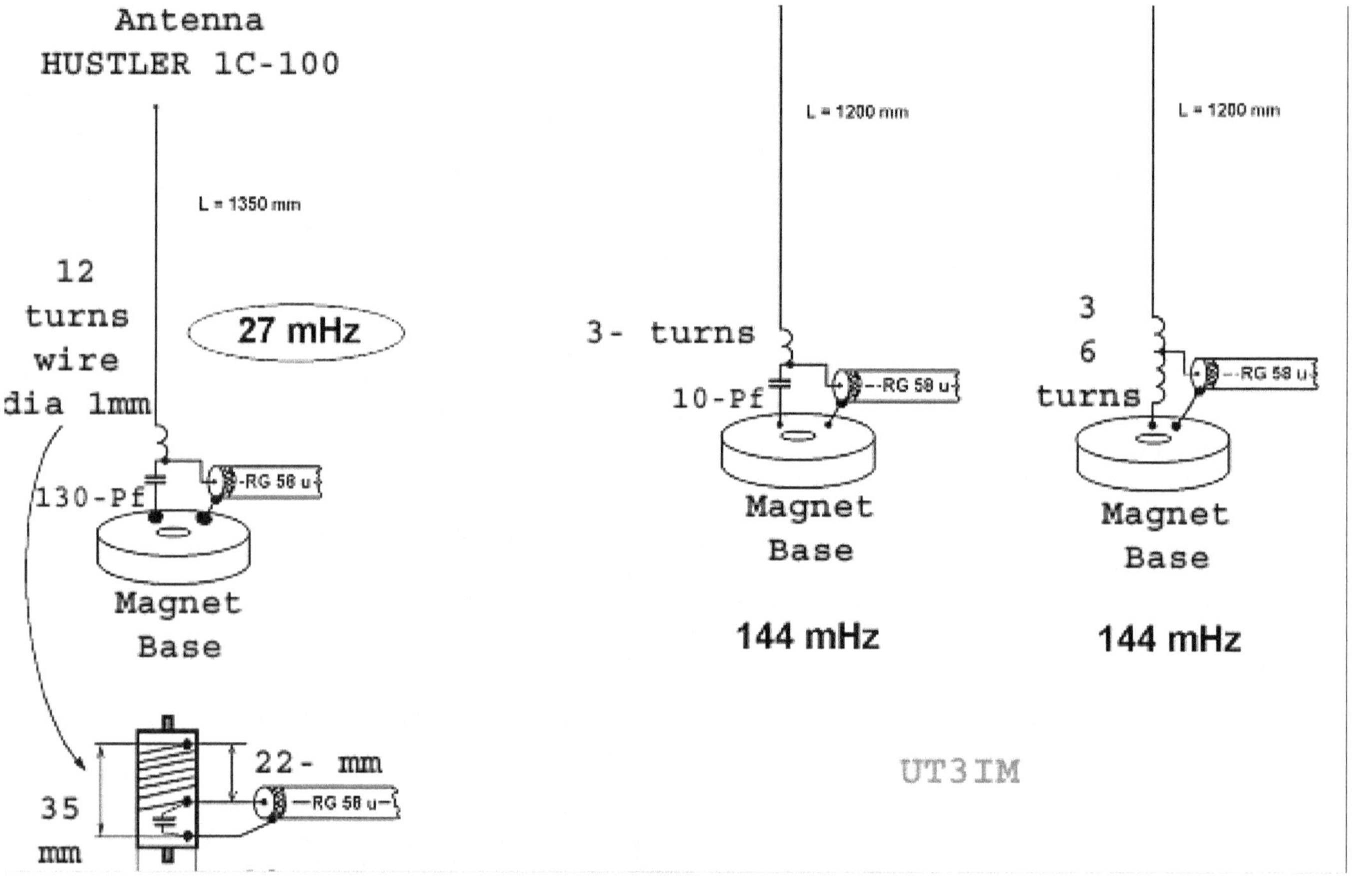

Figure 1 Conversion Auto CB-Antenna HUSTLER-1C-100 to Antenna for the 2- meter Band

www.cqham.ru

UB5UG Snake Antenna

Yuri Medinets, UB5UG

Note from I.G.: The Snake Antenna was very popular in the ex-USSR. The antenna came to ham from a page of paper with hand writer schematic. The schematic was introduced by Yuri, UB5UG. At first, the antenna was widely used at Ukraine then it came to other republics of the ex-USSR. It was a very simple antenna that very easy could be made from a coaxial cable. The antenna could be very easy redesign for other (as well for TV) bands.

As I remember, at the hand- writer of the antenna it was recommended do the antenna from a coaxial cable and the transformer 4:1 should be done from the same coaxial cable as used for antenna. In the ex-USSR it was widely used 75- Ohm coaxial cable. Antenna made from the cable works fine. At military application sometimes it was used 50- Ohm coaxial cable that sometimes it was possible to get. Antenna made from the 50- Ohm cable also worked fine.

Design of the antenna was very simple. It needs a suitable length of a wooden rod and a coaxial cable. Then with help electrical tape the coaxial cable was fastened to the wood rod. Transformer also it was fastened with the electrical tape to the wooden rod.

Opened parts of the coaxial cable were protected with help of usual plasticine. Antenna may be installed on the roof or at the balcony. At the 80-s (may be 90-s) the description of the antenna was published at Radio Magazine (Russia). However, I lost the issue of the magazine where the antenna was published.

73!
VA3ZNW

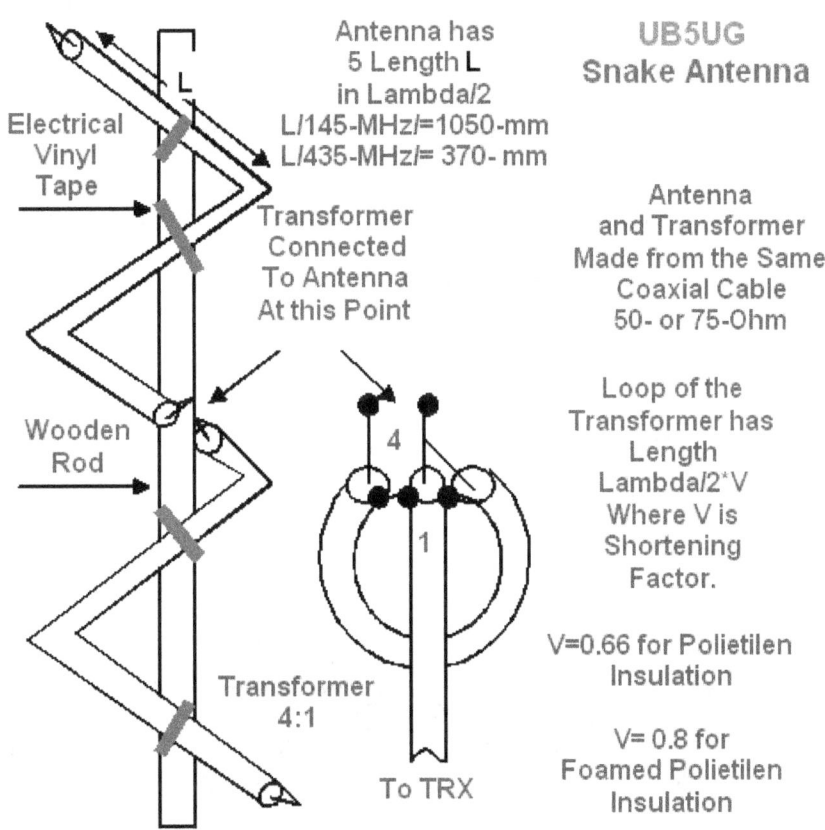

UB5UG Snake Antenna, as it was Pictured at Old Time Paper

Horizontal Antenna with Vertical Polarization for the 2- meter Band

Vasiliy Oleynik, RW4HX

At my vacation I guested my friends at their cottage near 40-km from the city. On the second day I decided to try my FT51. Oops, nobody can copy me when I transmitted on to the transceiver's rubber duck. So I need an antenna that could take the 40 km.
Experimenters with cottage- brew YAGI and Twin Loop antenna was failed. However I newer give up.
I found at the cottage:

10 meter aluminum wire in 4-mm diameter;
4 meters RG58 with connector for my transceiver;
An old TV- antenna – most valuable thing there was a broadband transformer 1:4.
Figure 1 shows antenna what I made from the stuff. Antenna was fastened by nails to wooden strip. The strip was placed under the cottage roof. It is a classical Chireix antenna that does not require strictly tuning and has broadband passband.

The antenna is exceeded all my expectations. Antenna provided very good communication at radius 40- 45- km also I could open two repeaters at distance more the 50- km.

The MMANA file of the Horizontal Antenna with Vertical Polarization for the 2- meter Band may be loaded: http: // www.antentop.org/018/rw4hw_018.htm

Figure 2 shows impedance of the antenna (antenna placed at 7- meter above the real ground). **Figure 3** shows SWR of the antenna (antenna placed at 7- meter above the real ground). **Figure 4** shows DD of the antenna (antenna placed at 7- meter above the real ground).

UR0GT is noticed, that the antenna would work fine at 10- meter Band. However at the band the antenna has input impedance near 50- Ohm. So, to do the antenna dual band a special ATU is needed. The Atu should work as transformer 1:4 at the 2- meter and as straight line at 10 meter.

Figure 5 shows impedance of the antenna (antenna placed at 7- meter above the real ground). **Figure 6** shows SWR of the antenna (antenna placed at 7- meter above the real ground). **Figure 7** shows DD of the antenna (antenna placed at 7- meter above the real ground).

Credit Line: www.cqham.ru

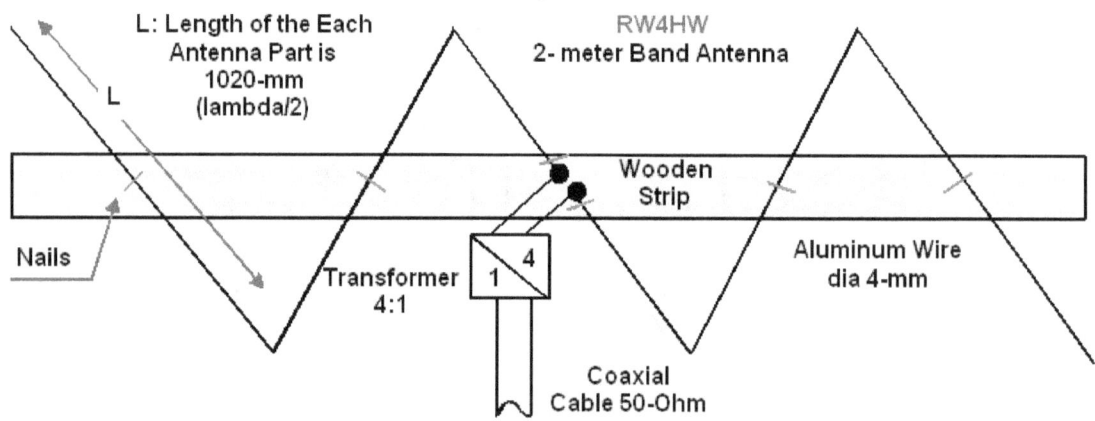

Figure 1 Horizontal Antenna with Vertical Polarization for the 2- meter Band

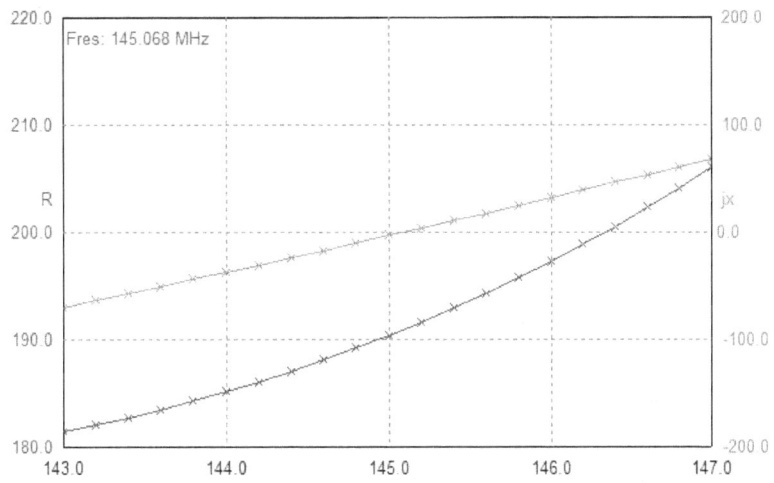

Figure 2 impedance of the Horizontal Antenna with Vertical Polarization at 2-meter Band

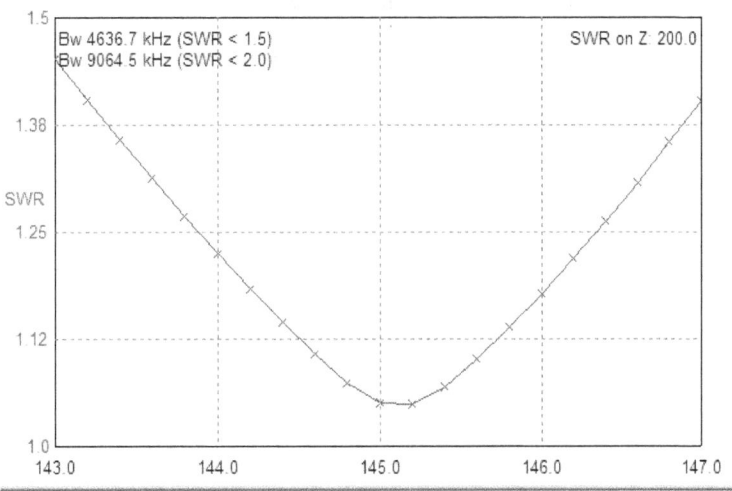

Figure 3 SWR of the Horizontal Antenna with Vertical Polarization at 2-meter Band

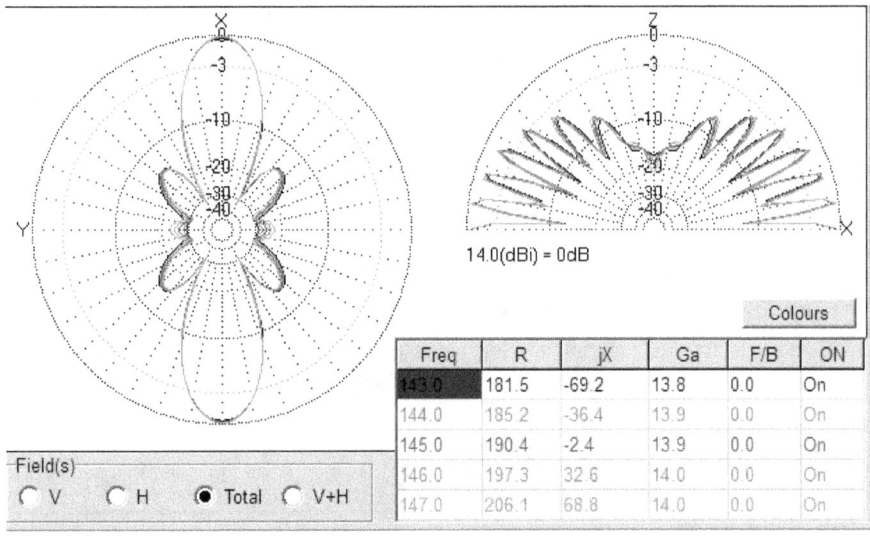

Figure 4 DD of the Horizontal Antenna with Vertical Polarization at 2-meter Band

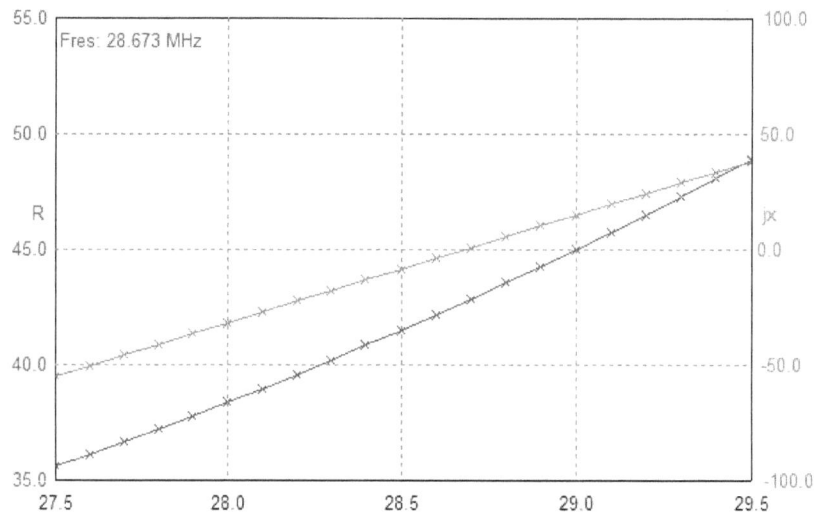

Figure 5 impedance of the Horizontal Antenna at 10- meter Band

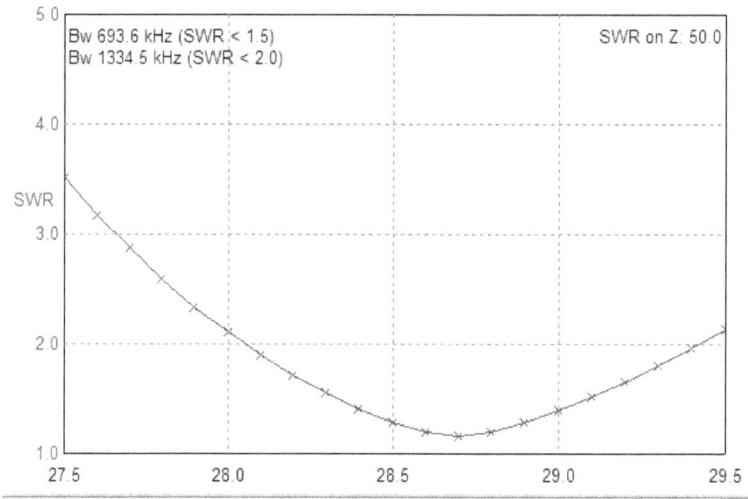

Figure 6 SWR of the Horizontal Antenna at 10- meter Band

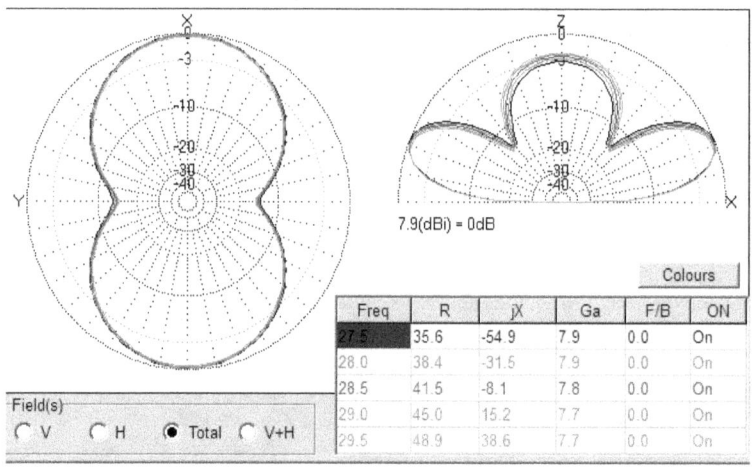

Figure 7 DD of the Horizontal Antenna at 10- meter Band

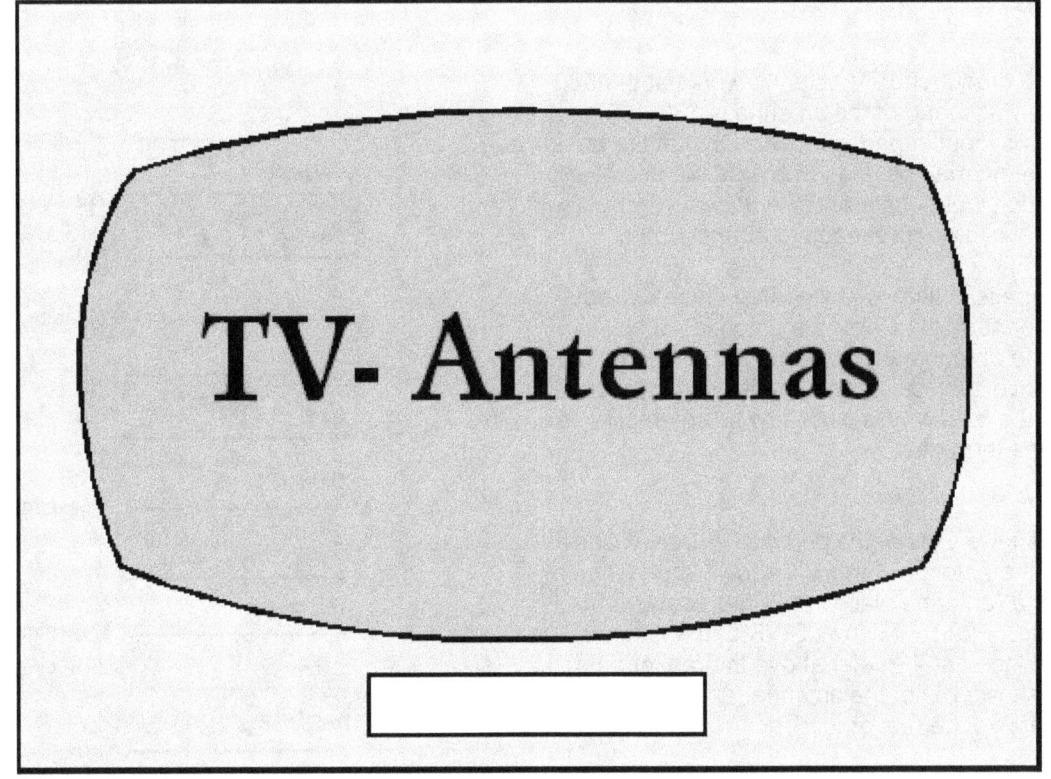

Broadband TV Antenna

The publication is devoted to the memory UR0GT.

Credit Line: Forum from: www.cqham.ru

By: Nikolay Kudryavchenko, UR0GT

The simple broadband TV antenna works at the 580- 760- MHz. Passband of the antenna is 180- MHz. Antenna has input impedance 300- Ohm at the pass band. Antenna may be used with antenna amplifier that has such input impedance. Antenna may be used with coaxial cable with broadband transformer.

The antenna is critical to any nearby metal subjects. They can destroy the DD of the antenna. Space in 50 cm near the antenna should be free from such metal or conductive subjects. If antenna is used for reception purposes the best way is place low noise amplifier at the antenna terminal.

Figure 1 shows view of the antenna. **Figure 2** shows design of the antenna. **Figure 3** shows impedance of the antenna (antenna placed at 7- meter above the real ground). **Figure 4** shows SWR of the antenna (antenna placed at 7- meter above the real ground). **Figure 5** shows DD of the antenna (antenna placed at 7- meter above the real ground).

The MMANA model of the Broadband TV Antenna may be loaded: http: // www.antentop.org/018/ur0gt_tv_018.htm

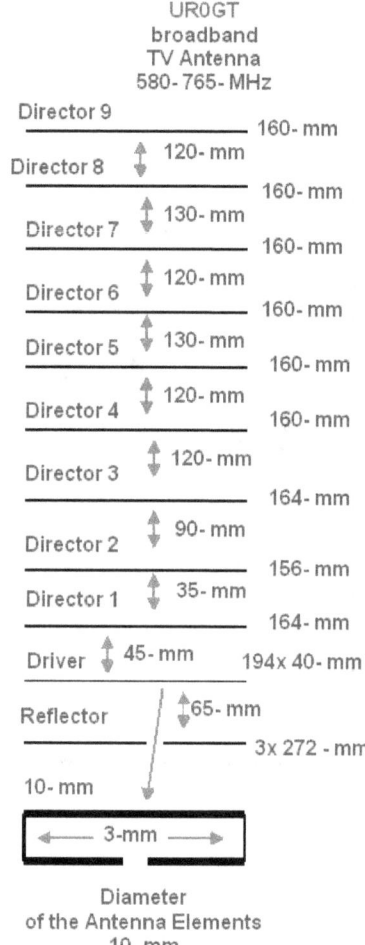

Figure 2 Design of the Broadband TV Antenna

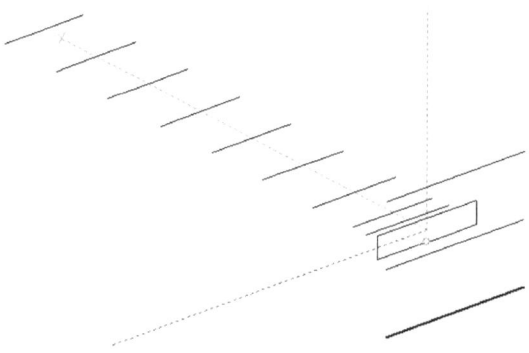

Figure 1 View of the Broadband TV Antenna

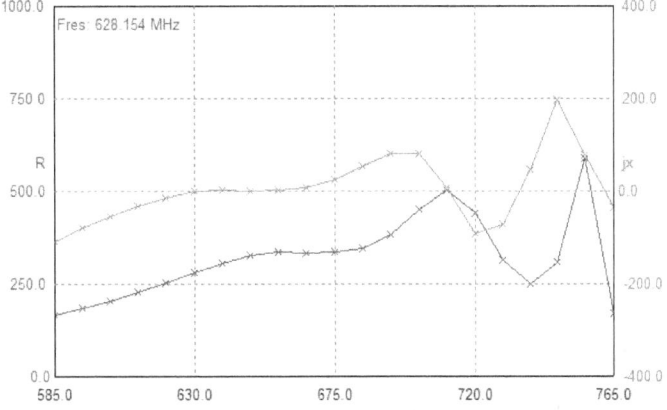

Figure 3 impedance of the Broadband TV Antenna

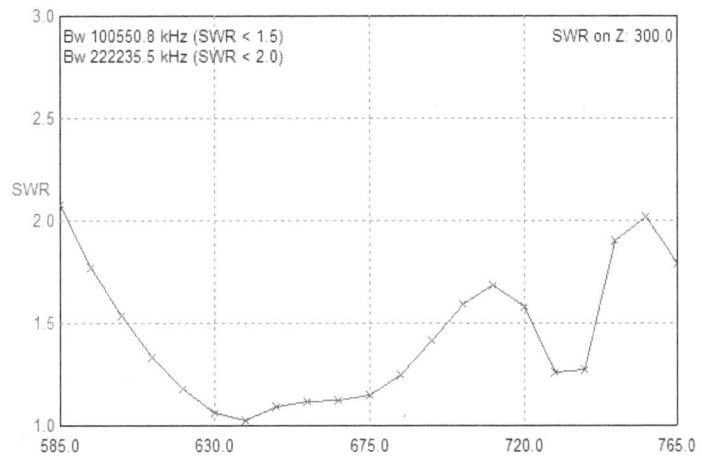

Figure 4 SWR of the Broadband TV Antenna

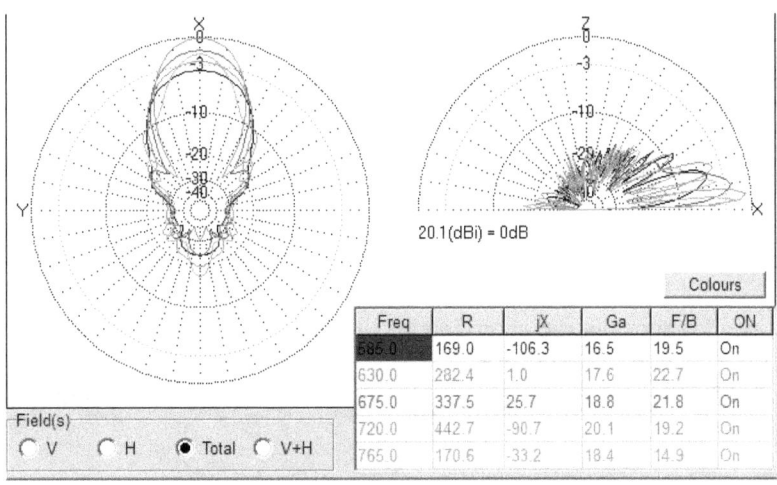

Figure 5 DD of the Broadband TV Antenna

73 Nick

Chireix- Mesny TV Antenna

The publication is devoted to the memory R0GT.

Credit Line: Forum from:
www.cqham.ru

By: Nikolay Kudryavchenko, UR0GT

The simple broadband TV antenna works at the 615- 765- MHz. Antenna has input impedance 300- Ohm at the pass band. Antenna may be used with antenna amplifier that has such input impedance.

The antenna is a variant of the famous Chireix- Mesny Antenna.

Antenna may be used with coaxial cable with broadband transformer. switched to the TV 150- MHz.

Figure 1 shows design of the antenna. **Figure 2** shows impedance of the antenna (antenna placed at 7- meter above the real ground). **Figure 3** shows SWR of the antenna (antenna placed at 7- meter above the real ground). **Figure 4** shows DD of the antenna (antenna placed at 7- meter above the real ground).

The MMANA model of the Chireix- Mesny TV Antenna may be loaded: http: //
www.antentop.org/018/chireix_018.htm

Note I.G.: Chireix- Mesny Antenna was designed in France by Henri Chireix, Chief Engineer of the Societe Francaise Radiotelectrique, and Rene Mesny, Professor of Hydrography in the French Navy. Papers on the antenna were published (in different variations) in the 1926- 1928s. Patent H. Chireix: French Patent # 216,757, filed Mar. 10, 1926.

Antenna originally was used for directive radiation and reception at short waves. Lately the antenna was widely used at VHF- UHF waves.

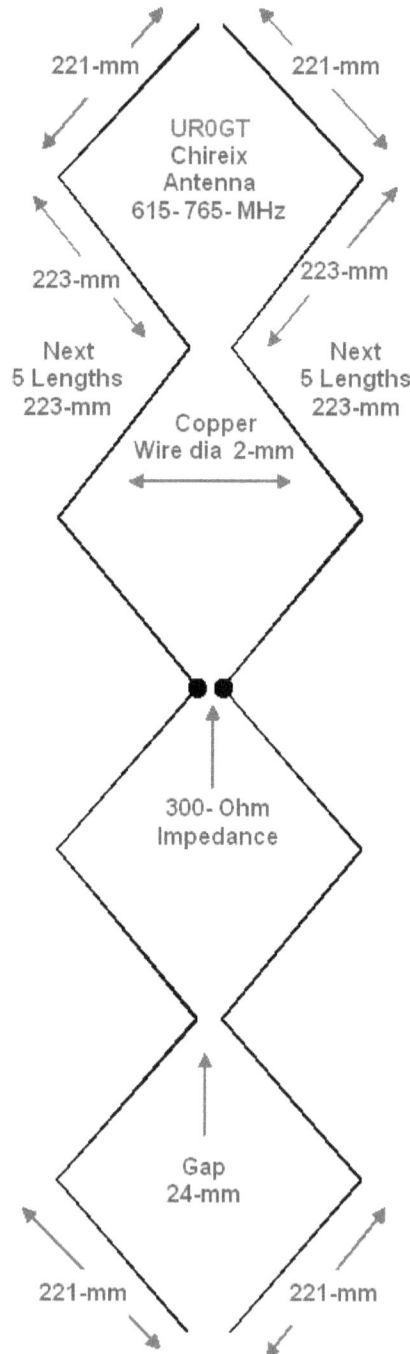

Figure 1 Design of the Chireix- Mesny TV Antenna

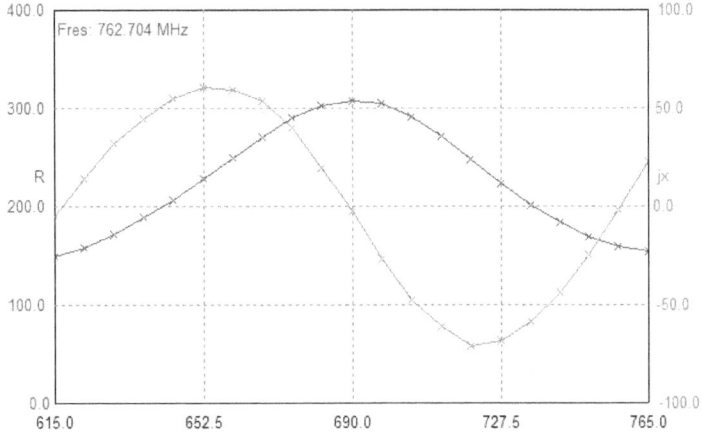

Figure 2 impedance of the Chireix- Mesny TV Antenna

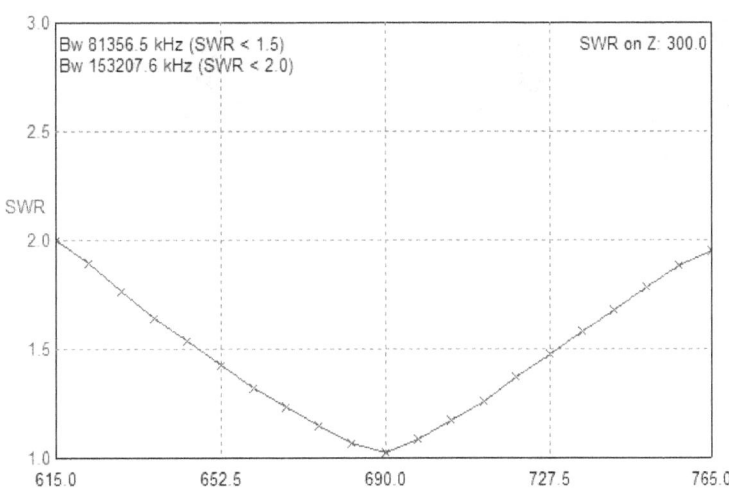

Figure 3 SWR of the Chireix- Mesny TV Antenna

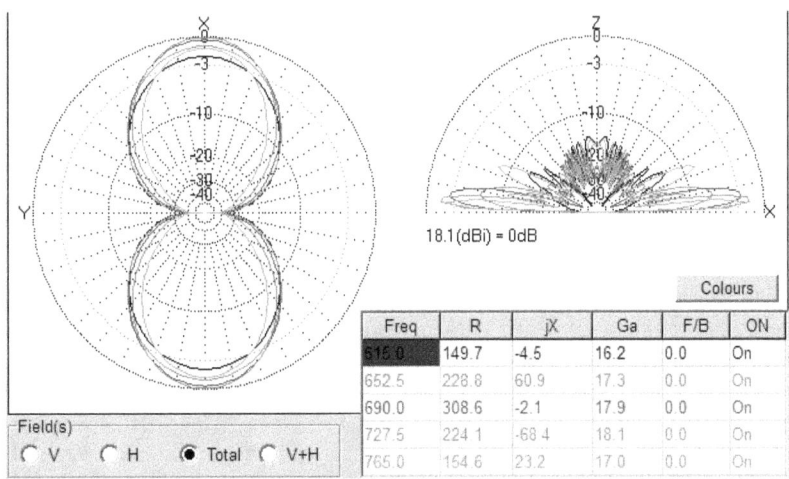

Figure 4 DD of the Chireix- Mesny TV Antenna

73 Nick

Useful Pieces

Tube Socket from Surplus VHF Resonator

Robert Akopov, UN7RX

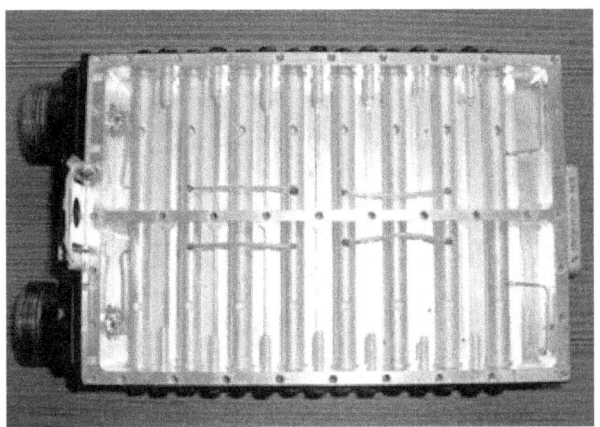

It was used surplus VHF device. VHF resonators were at bottom side,

at the front side there was a board with VHF circuit. The resonators made from silvered bronze.

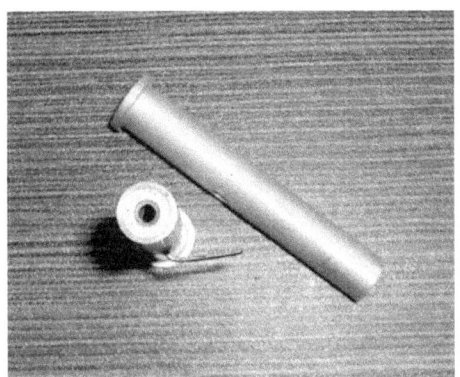

Resonators were removed. Then central hole was drilled to diameter of the tube's pin.

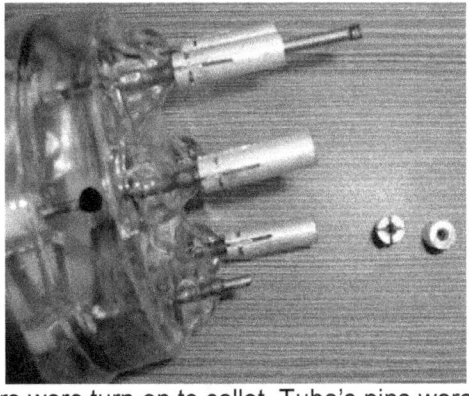

Cylinders were turn on to collet. Tube's pins were polished to smooth inserting into the holders.

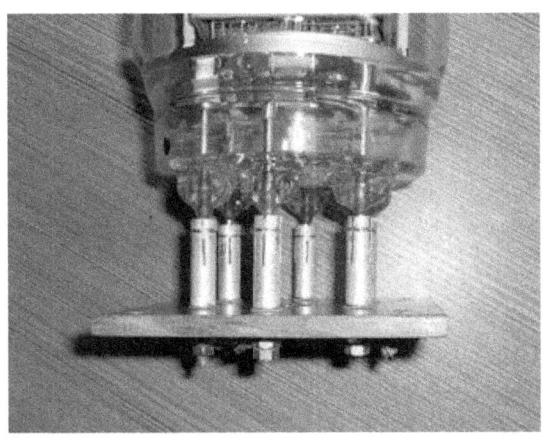

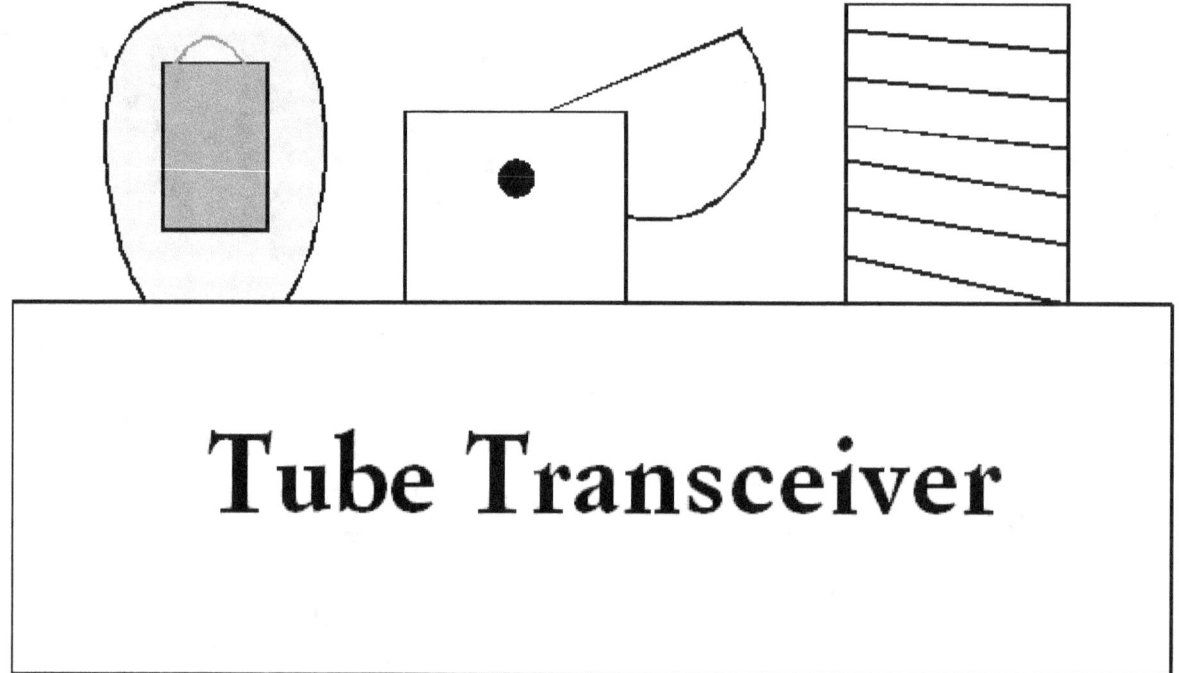

www.antentop.org

Regenerative Transceiver for 160- meter Band

By Georgiy Gorelashvili, 4L1G

There is experimental AM transceiver that may be used at local communication. Practically any pentode would work at the circuit. Frequency of the transmitter is not stable because antenna is switched on directly to the oscillator's inductor. Regime of the regenerative receiver could not be optimal because of high coupling of the antenna with receiver's inductor. However, at small antenna length – in 2… 5 meter and small distances the transceiver is quite well for experimental work. Transceiver may be retuned for high bands however stable of operation at the bands would be low. Antenna length at the high bands should be decreased to 1 meter.

Parts List.

Microphone: Carbon microphone from old telephone set.

Transformer 1 and 2: An audio transformer from an old tube radio receiver would work fine

Relay: Any suitable Relay with 4- groups contacts.

L for 160-m: 50-turns, wire 0.5-mm (AWG-24). Tap from 20s turn from cold end. Diameter 15… 18-mm.

L for 40-m: 20-turns, wire 0.7-mm (AWG-21). Tap from 6s turn from cold end. Diameter 30-mm. Coiled with gap between turns 2 mm.

L for 20-m: 10-turns, wire 0.7-mm (AWG-21). Tap from 4s turn from cold end. Diameter 30-mm. Coiled with gap between turns 2 mm.

L for 10-m: 5-turns, wire 0.7-mm (AWG-21). Tap from 2s turn from cold end. Diameter 30-mm. Coiled with gap between turns 2 mm.

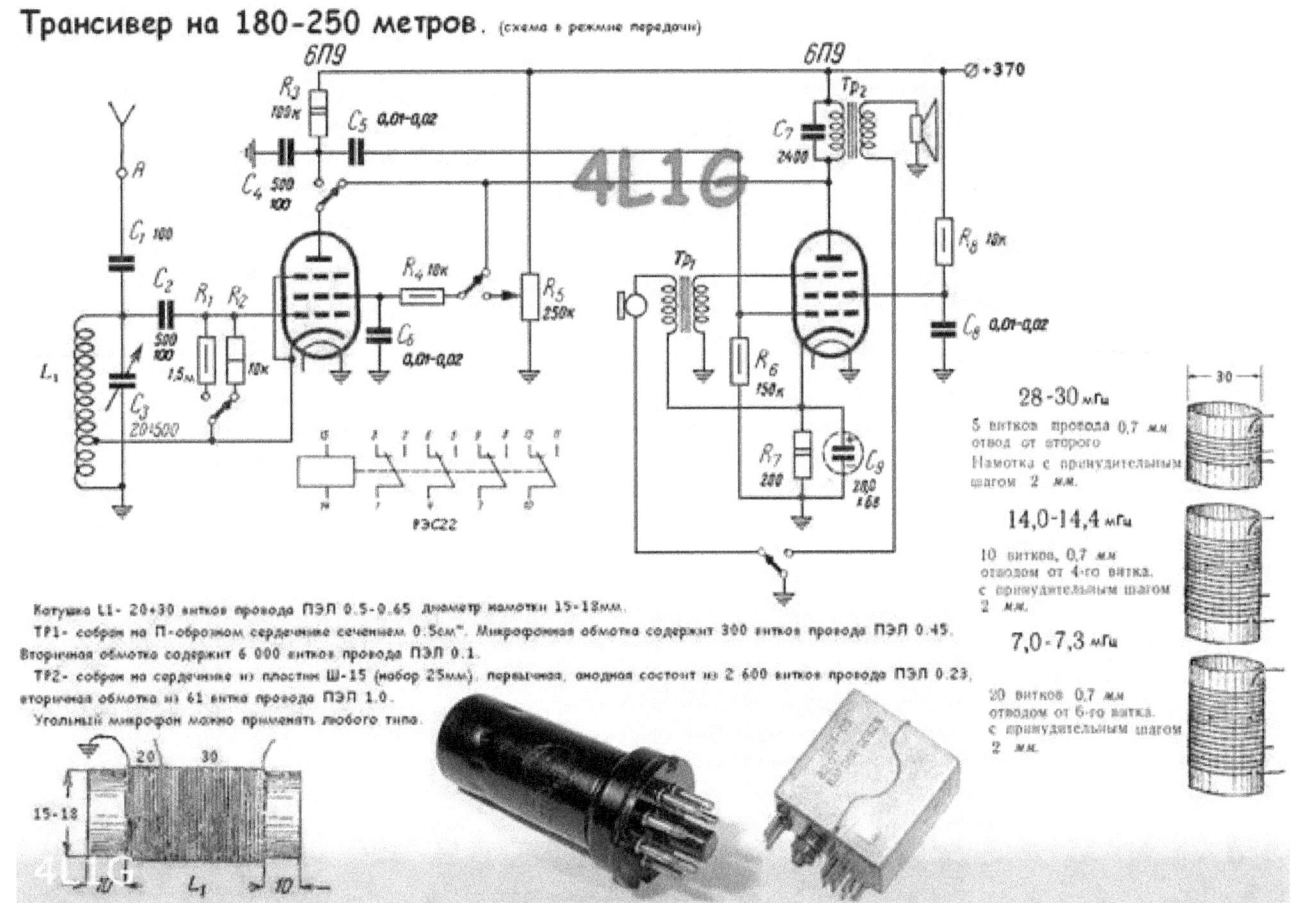

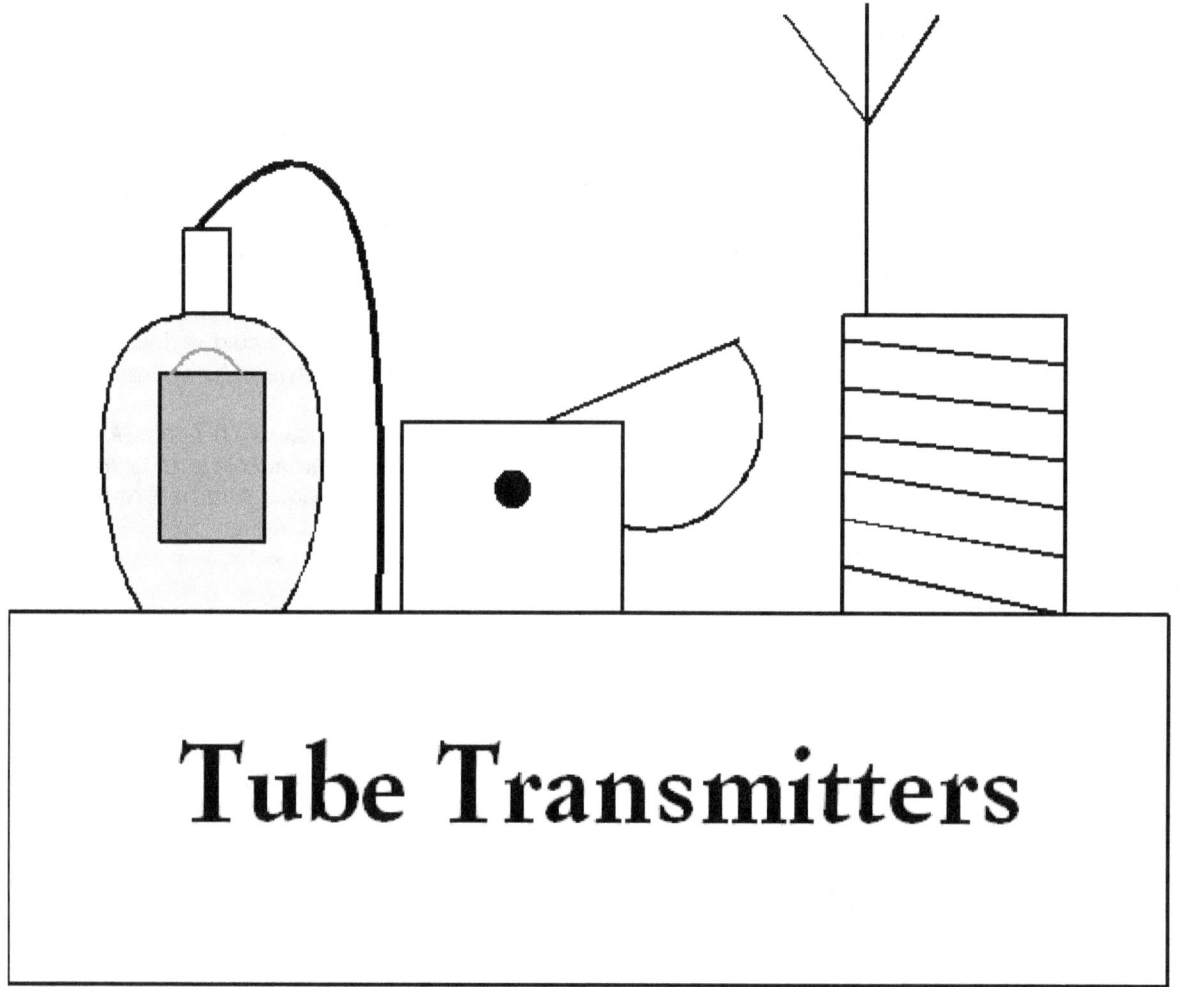

Retro AM Transmitter for 160-m

By Georgiy Gorelashvili, 4L1G

The schematic (in different variations) was very popular in ex-USSR. There is possible use any powerful pentode with tap from the third grid (for modulation). Of course, frequency of the transmitter is not stable but it is quite possible use for experimental purposes.

Parts List:

Microphone: Carbon microphone from old telephone set.

Transformer: An audio transformer from an old tube radio receiver would work fine.

L1: 58 turns, wire 0.5-mm (AWG-24), diameter 30-mm, tap made from 13s turn from cold end.

L2: 38 turns, wire 0.5-mm (AWG-24), diameter 40-mm.

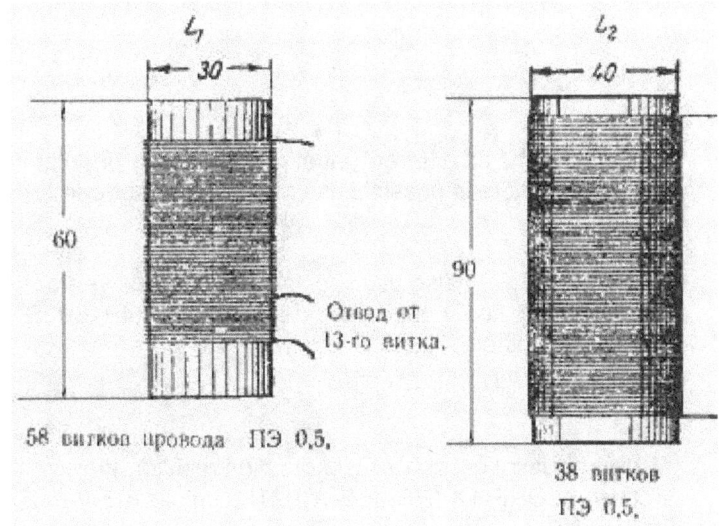

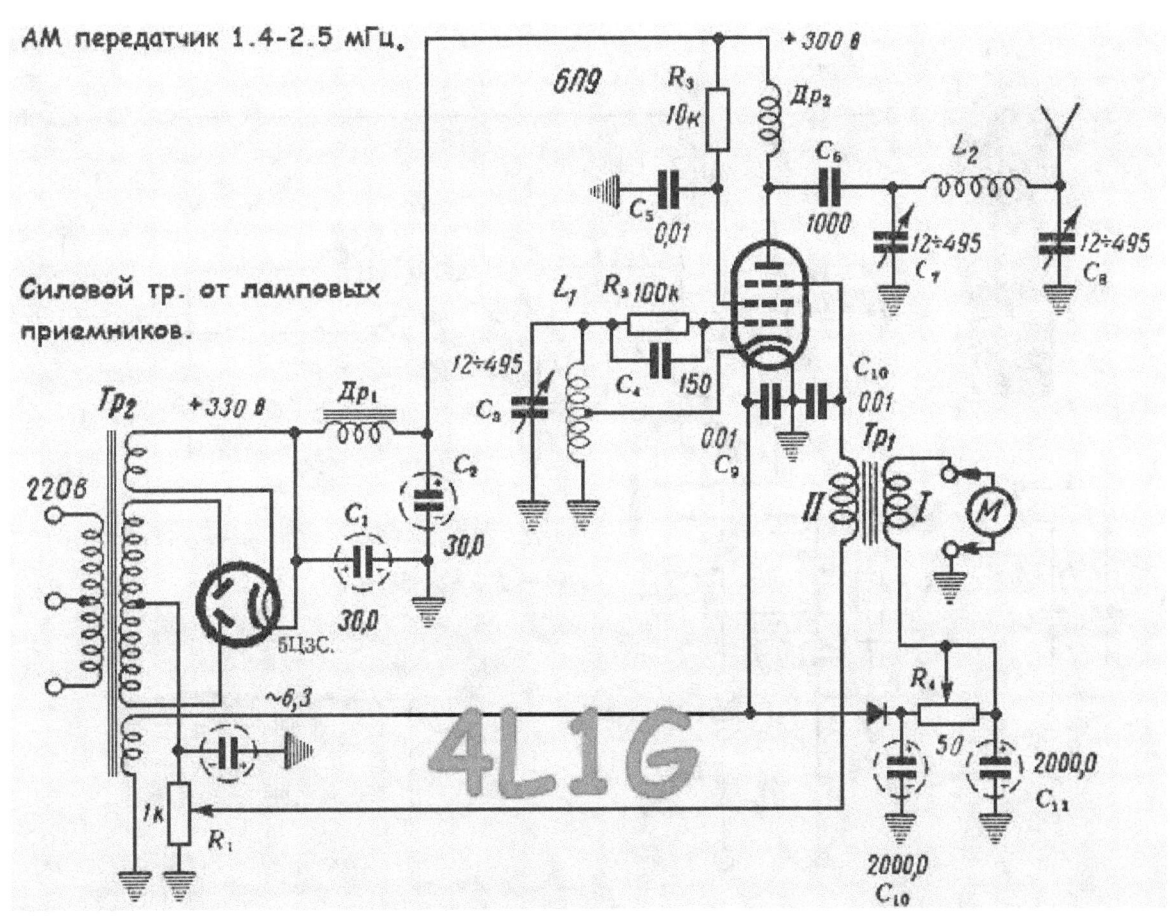

CW Tube Transmitter for 40 and 80- meter Band

By Georgiy Gorelashvili, 4L1G

The schematic (in different variations) was very popular in ex-USSR. There is possible use any powerful tetrode for the rig.

(*Note from I.G.:* I also used such rig in 70s- 80s years. The rig works fine. I used pentode 6P15P, 6P14P, 6P3S at the transmitter)

Parts List.

Bulb at quartz resonator is protection resistor. It may not use at QRP-TX (NEAR 5-Wtt). The bulb is 6.3-V- 50-mA.

Bulb at antenna circuit: 3.5v- 0.15…0.5-A. Amperage depends on antenna impedance.

L: 30 turns, wire 0.9- mm (AWG-19), coiled with small gap between turns, tap from the middle, diameter 40- mm.

RFC1: About 150-microH.

RFC2: About 100-microH.

RFC3: 20 turns on resistor 1-Wtt. Wire 0.2-mm (AWG-35)

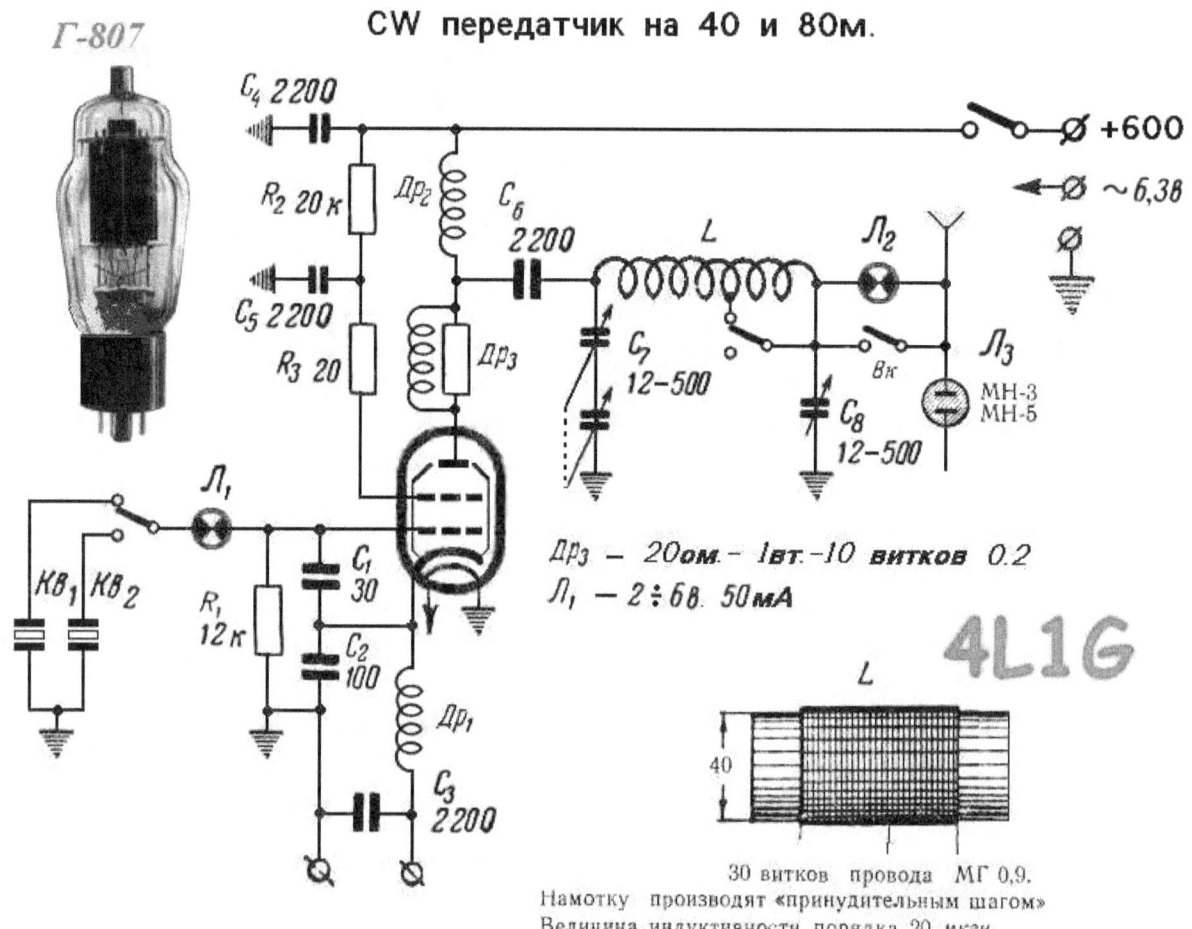

Simple AM Tube Transmitter for 1.5- 3.5- MHz

By Georgiy Gorelashvili, 4L1G

4L1G

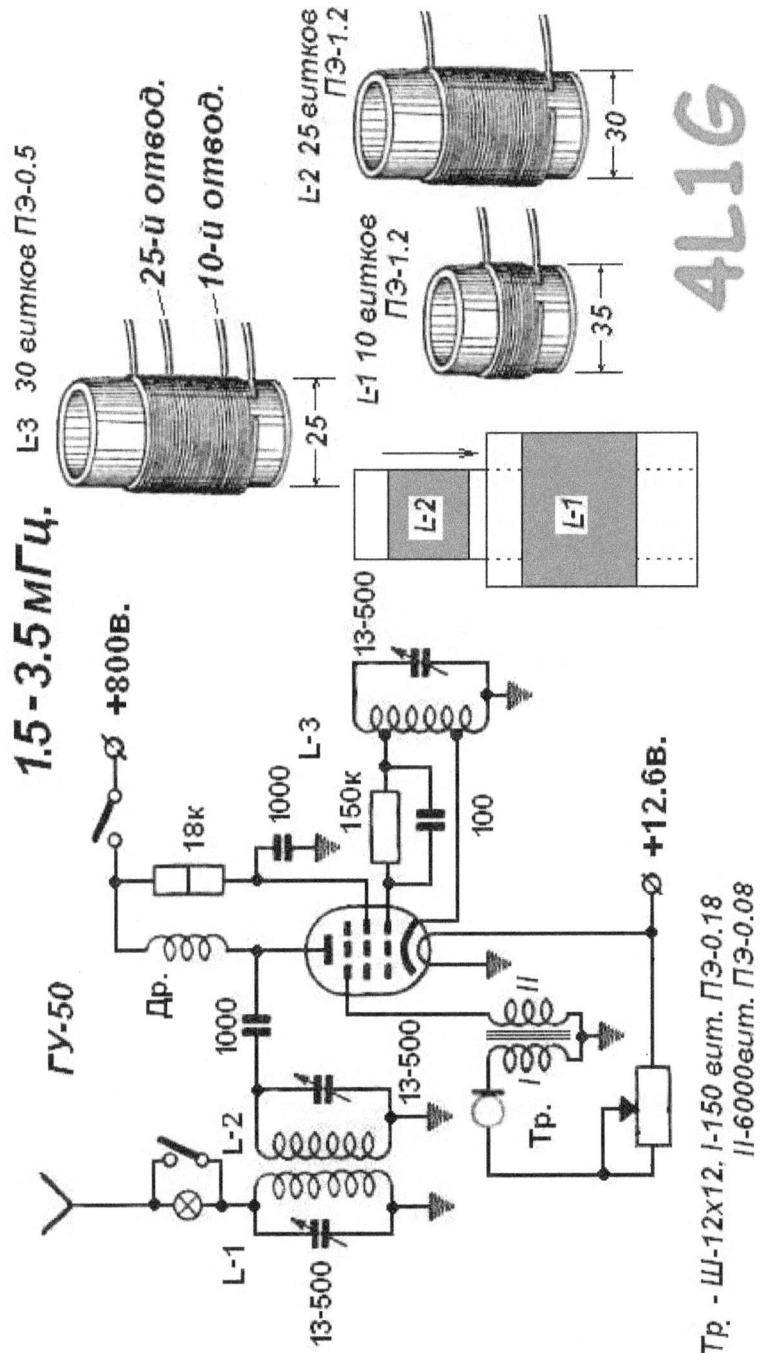

The schematic (in different variations) was very popular in ex-USSR. There is possible use any powerful pentode with tap from the third grid (for modulation). Of course, frequency of the transmitter is not stable but it is quite possible use for experimental purposes.

Parts List:

Microphone: Carbon microphone from old telephone set.

Transformer: Best choice is to use a special transformer from surplus military radio. Transformer is hard to make because the second winding contains 6000 turns of wire 0.08- mm (39- AWG). However, an audio transformer from an old tube radio receiver would work fine.

Bulb at antenna circuit: 3.5v- 0.15...0.5- A. Amperage depends on antenna impedance.

L1: 10 turns, wire 1.2- mm (AWG-16), diameter 35-mm.

L2: 25 turns, wire 1.2-mm (AWG-16), diameter 30-mm.

L3: 30-turns, wire 0.5-mm (AWG-24) Tap from 25s and 10s turn. Diameter 25-mm

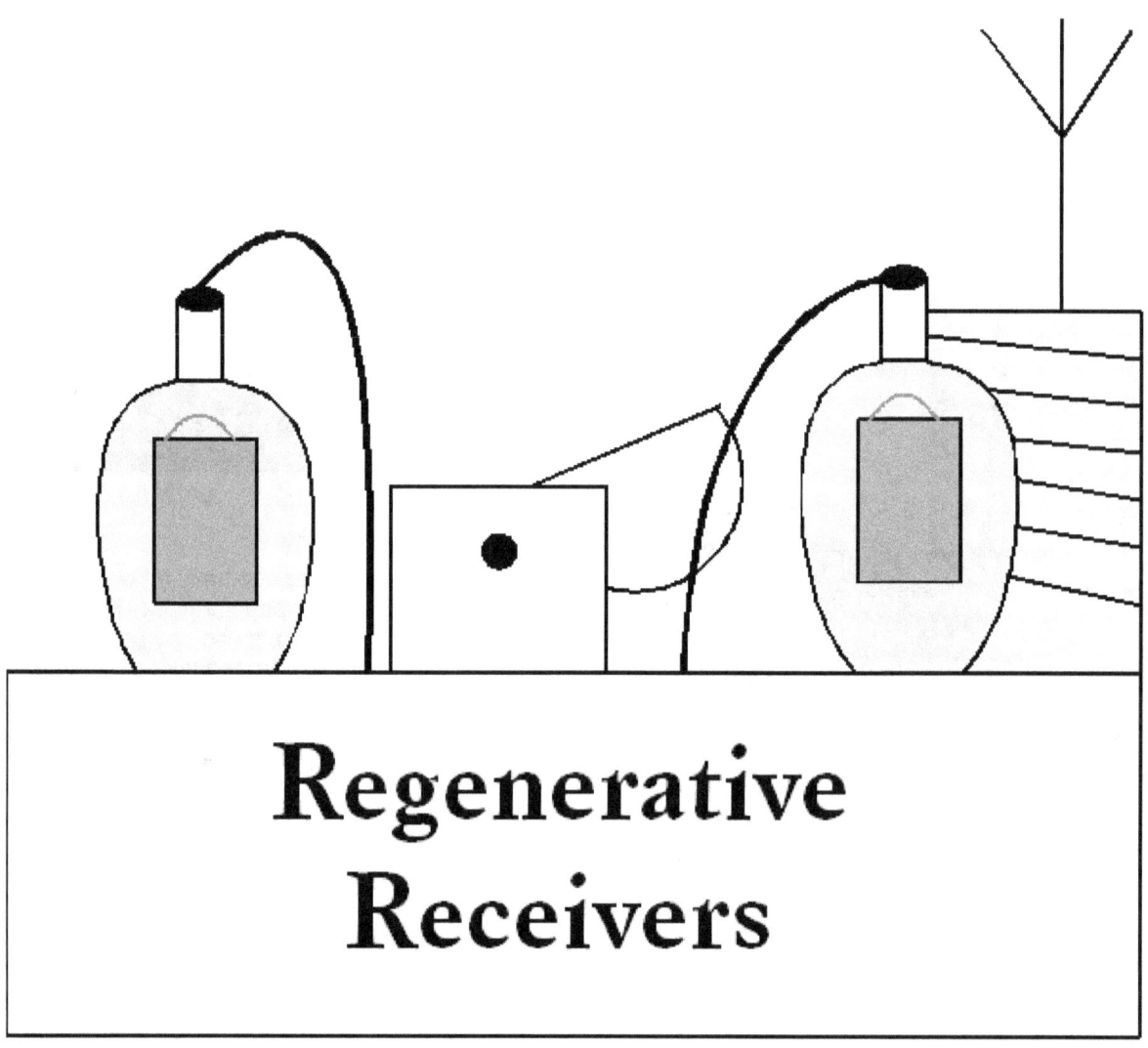

Simple MW- HF- Regenerative Receiver

Seiji
Credit Line: http://www.cqham.ru/forum/showthread.php?t=14624&page=435

Simple MW- HF Regenerative receiver has two independently stages- one for MW-Band another one for HF-Band. HF Band is covered 6.3- 9.0- MHz. One two-section variable capacitor (C26) is used for both sections. HF Section has fine tuning- there are VD1 and R2. R6 and R8 is regeneration control for MW and HF- Band. R22 is Audio Gain.

LED1 shows tuning on the station. Circuit on VT5, VT6, VD2 and VD3 is Automatic Gain Control. Diodes VD2 and VD3 are germanium ones. Speaker is 8- Ohm/1- Wtt. **Figure 1** shows schematic of the receiver. **Figure 2** shows inductors for MW and HF- Band. Tap at MW- Band made from 1 turn from cold end.

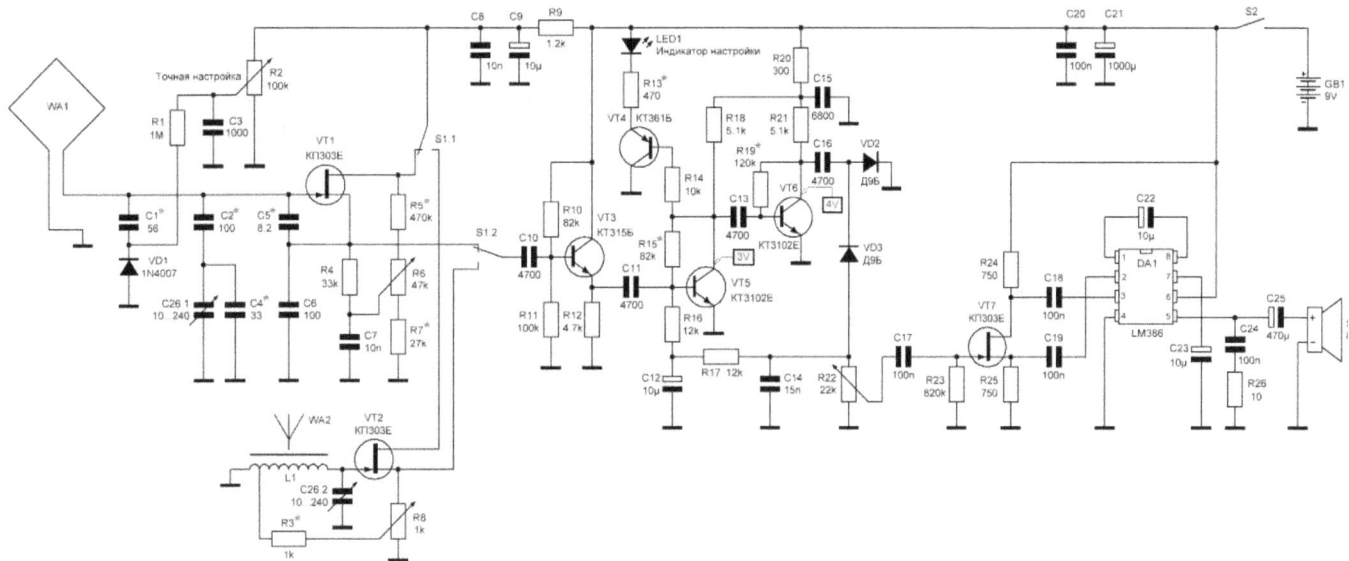

Figure 1 Schematic of the Simple MW- HF- Regenerative Receiver

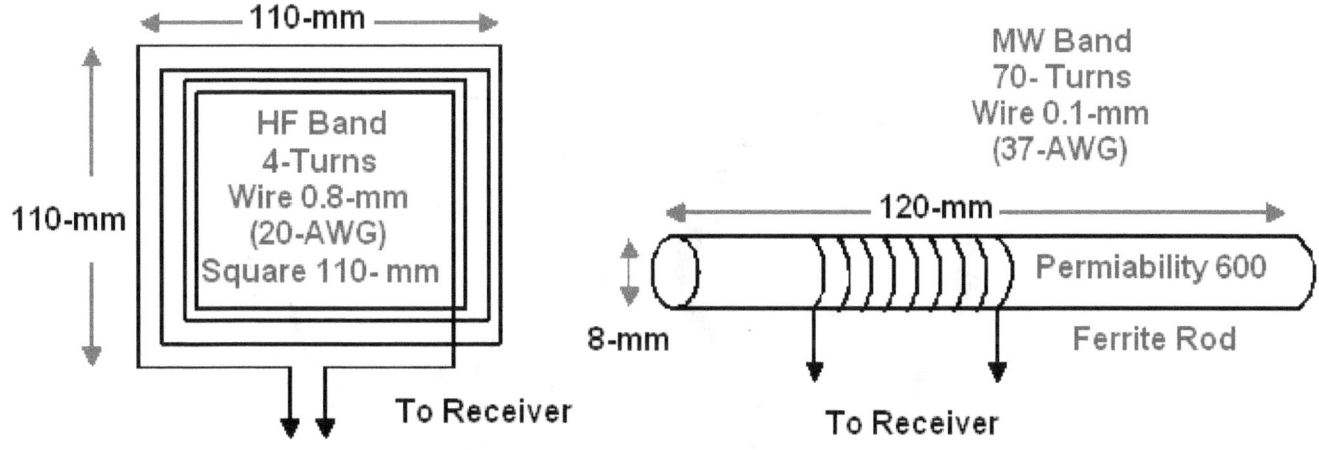

Figure 2 Inductors for MW and HF- Band

Simple Regenerator Receiver with Loop Antenna

Aleksandr Bulanenko, UA6AAK

It is next generation of simple battery powered regenerative receiver. The receiver used to the input inductor as antenna. It was made two inductors, one for 3.3- 14.6- MHz another one for 4.0- 19.0- MHz.

Figure 1 shows schematic of the receiver. **Figure 2** shows design of the Loop inductors of the receiver. Audio files of the receiver you may find at youtube.

Video plus Audio: https://www.youtube.com/watch?feature=player_embedded&v=T9qw_UBhRIM

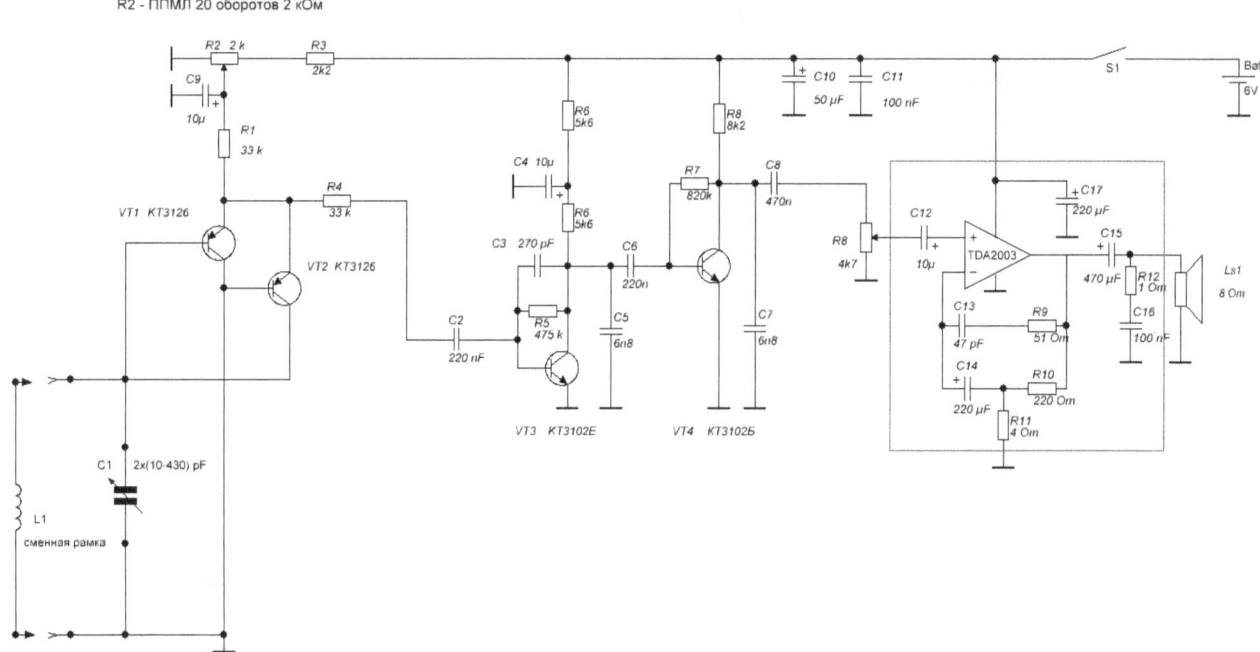

Figure 1 Schematic of Simple Regenerator Receiver with Loop Antenna

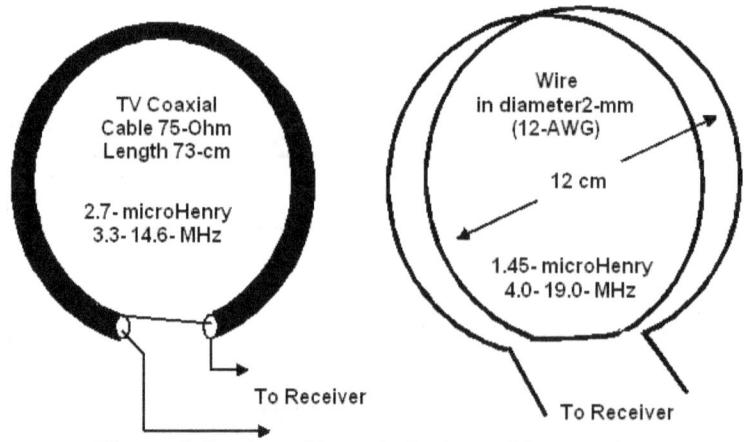

Figure 2 Design of Loop inductors of the receiver

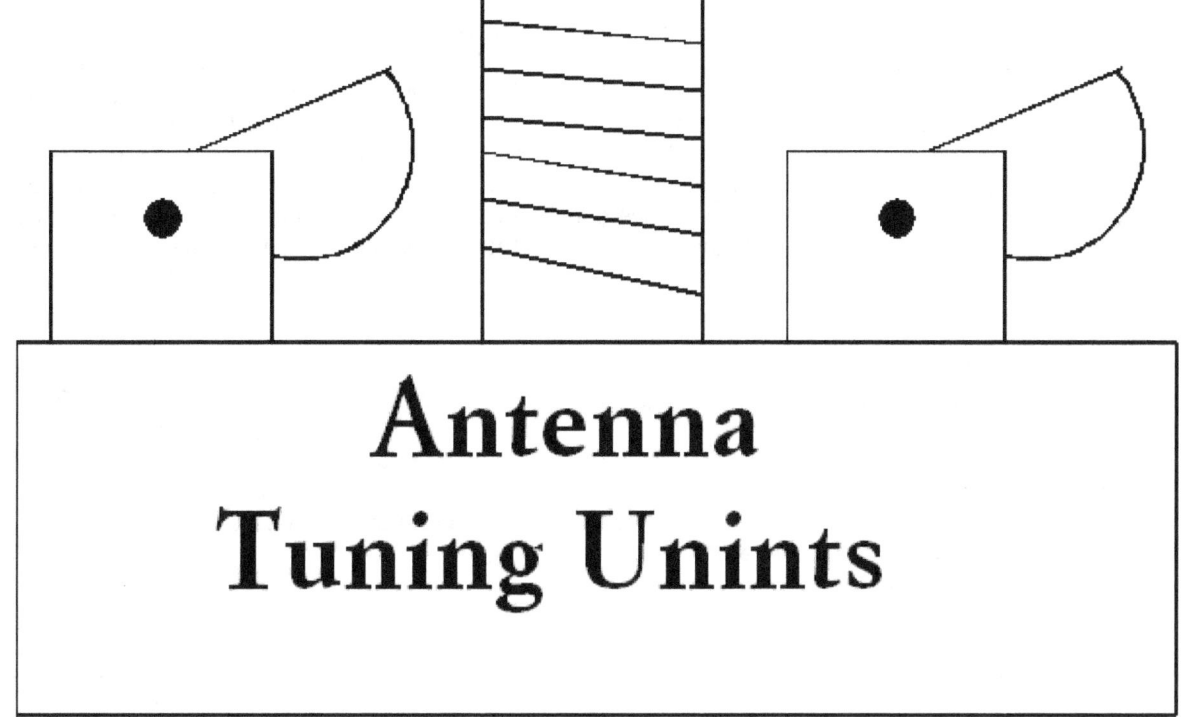

Simple HF ATU on Lengths of Coaxial Cable

Igor Grigorov, va3znw

The ATU was made by me in far 80s. It was may be a simplest ATU what I made ever. It contains only one rotary switch and rolls of a coaxial cable. But the ATU works very well. The ATU has only one lack- sizes. Sizes of the ATU are not small. Below there are several words to the theoretical base of the ATU.

For those who know the Smith Chart the principle of operation of the ATU is not a secret. Using Smith Chart we can find how impedance of antenna system is changed along a transmission line. Based on this we can find optimal length of the transmission line and place(s) on transmission line where we can install stub(s) to eliminate the reactance in the line. Of course, the description is too simple and GOOGLE helps those who want to know more about Smith Chart and how we could match antenna impedance with our transmitter using only transmission line of definitely length with stubs.

For described here matching unit it means: let's turn on our antenna through line with variable length and find when SWR at our transmitter would be best. It is easy. It is simple. It works in most cases.

Yes, at some cases the ATU does not work or work not good but only in some cases… **Figure 1** shows schematic of the ATU.

The ATU contains 11 lengths of 50-Ohm coaxial cable. First length is 1 meter long, second one is 2 meter long, third one is 3 meter long, and so on, next one has length in 1 meter longer the previously one. The lengths of coaxial cable are connected to a four-pole 11-position rotary switch S1. So with help of the S1 you may choose the length of the transmission line from transmitter to antenna system. And if you are lucky (you will be lucky!) match the transmitter with existing antenna system. The matching device works fine at HF- Range 3.5- 30.0- MHz.

If you wish use the device only at 7.0- 30.0- MHz the lengths of the coaxial cable should have step 0.5- m. First one should have 0.5- m length, the second 1.0- meter length, the third one 1.5- meter length and so on. **Figure 2** shows the design of the ATU.

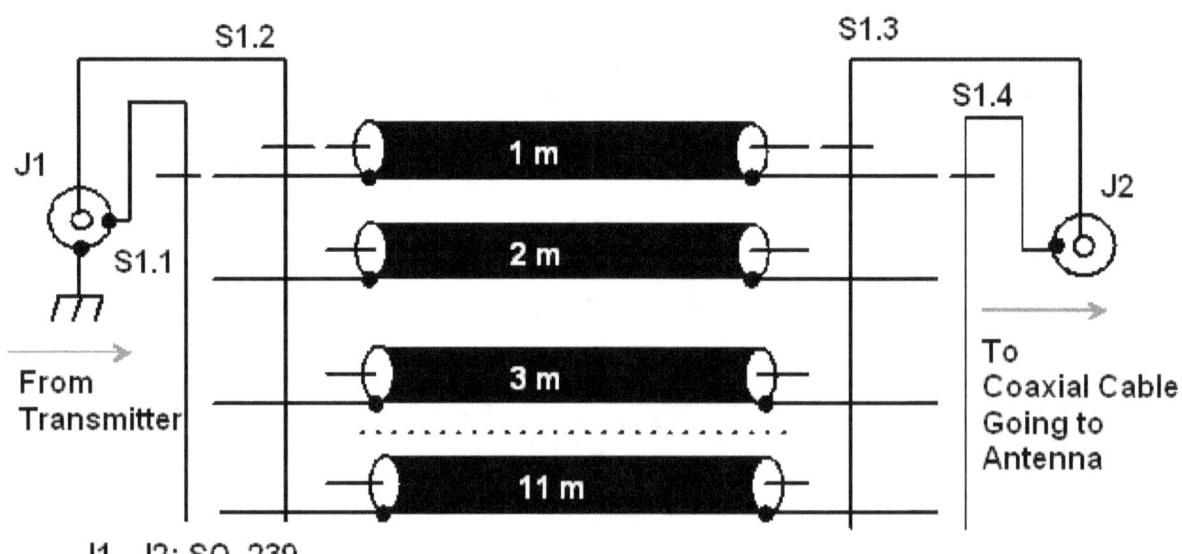

Figure 1 Simple HF ATU on lengths of Coaxial Cable

Note: Pay attention that "ground" of the socket J2 is not connected to the ATU case. The socket is placed on a dielectric plate (it was used a piece of PCB without foil).

The design was very simple. A big metal box from unknown surplus device was used for the ATU. Lengths of the coaxial cable was coiled and then dressed on to a plastic tube. It was used plastic water pipe (something like 1… 2-inch OD).

The ATU is very simple to use. Just connect the ATU between transmitter and antenna system. Then rotate S1 on to minimum SWR. **Figure 3** shows connection of the ATU. It is possible to use the transceiver's internal SWR–meter or an external one. Do not rotate the S1 when transceiver is in transmission mode. S1 breaks the transmission line so it may cause high SWR. Go to receiving mode, switch the length of the coaxial cable, go to transmission mode and check SWR.

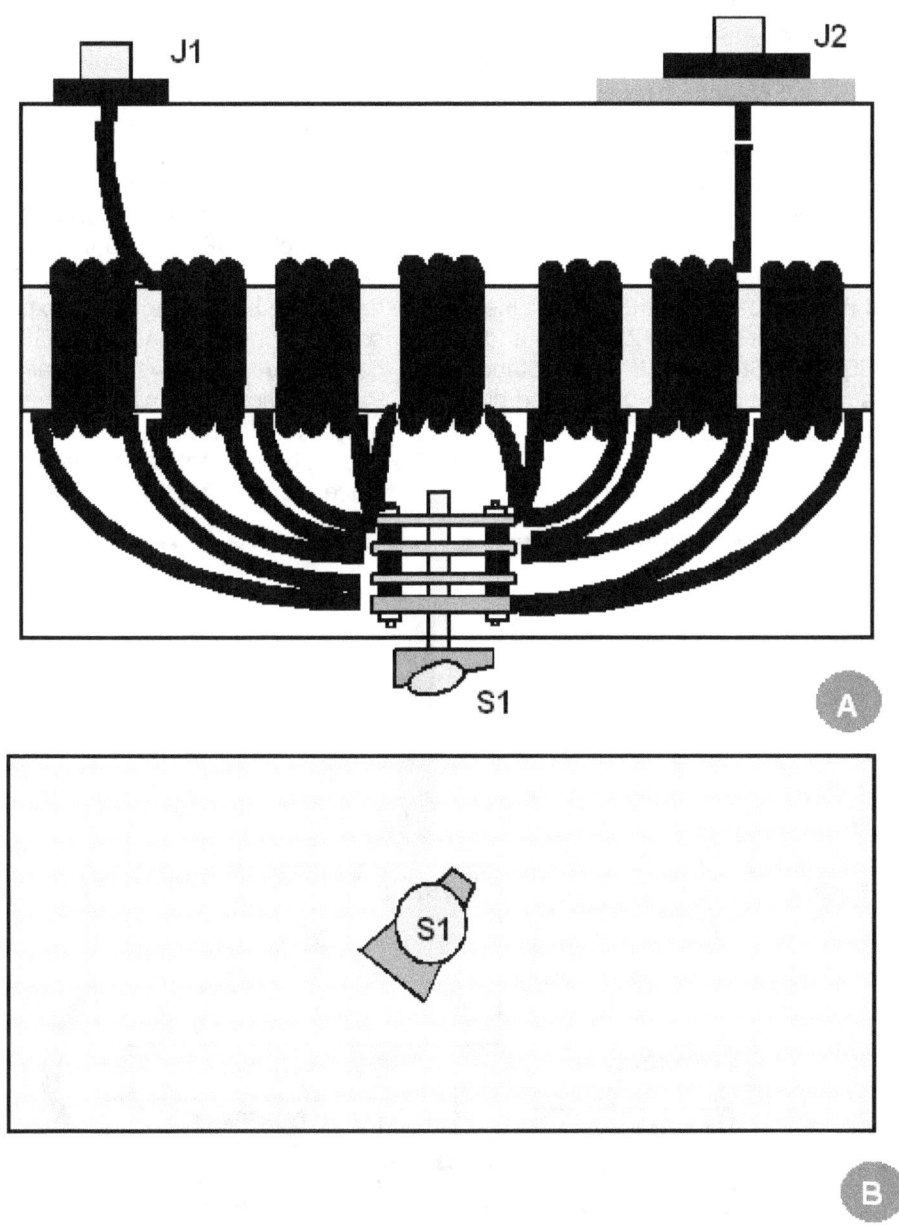

Figure 2 Design of the Simple HF ATU on lengths of Coaxial Cable

However, you may rotate the S1 in receiving mode and tune on to maxima reception. You need begin the tuning from the lowest length of the length of the coaxial cable. Then turn on the next length of the coaxial cable and check SWR.

On my memory it was very effectively ATU that could match lots of my experimental antennas (be truth – antenna system- antenna plus coaxial cable going to my transmitter) that I just connected to the coaxial cable placed on the roof. One of the days I decided to rework the ATU to get more efficiency. Any lover of the Smith Chart and matching of the antenna by length of the transmission line straight away could understand how the next ATU works. For those, who do not care about the theory I include some simple explanation. **Figure 4** illustrates the explanation.

For example, there is an antenna with impedance, let's say, 500 Ohm and Reactance minus 100 Ohm. Line L3 transform the impedance, let's say, to 200 Ohm and Reactance Plus 300-Ohm. Stub L2 kills the reactance. The stub, depends on the length, may be opened or closed. Then line L1 transform the pure 200- Ohm to 50- Ohm at the transmitter terminal.

Of course, it is very simple explanation, in the real life, is required work with the Smith Chart to find the gold length of the L1, L2 and L3. I have a doubt that in real life somebody will do the theoretical simulation at amateur station. But we may try it in practical way. May be we would be lucky and maybe we can do it with our ATU.

At the design of our ATU we have all components that are at **Figure 4**. Line L3- it is coaxial going from the antenna to the ATU. Line L1- it is variable length of the coaxial cable that we connect between transmitter and antenna system. Line L2- it is row of length of not used coaxial cables that are sitting inside ATU. What can we do- just connect the unused cable in to terminal ATU- antenna system. Then play! Chose length from transmitter to antenna system, then connect the stub and check SWR, and again try another connection to find the low SWR at transmitter terminal. At first sight is hard to do but having some experience (and maybe some theoretical base- GOOGLE helps you) it is not so hard. **Figure 5** shows schematic of the modified ATU. Pay attention that Switch S2 has one empty position (physically I removed stopper from the switch) when no one stub does not connected to the ATU. Switch S2 was placed under switch S1.

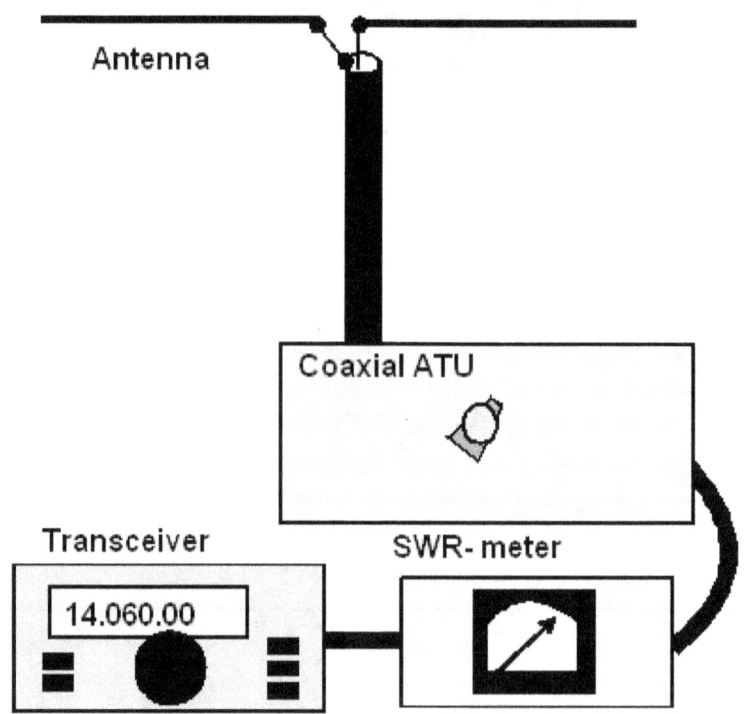

Figure 3 Connection of the Simple HF ATU on lengths of Coaxial Cable

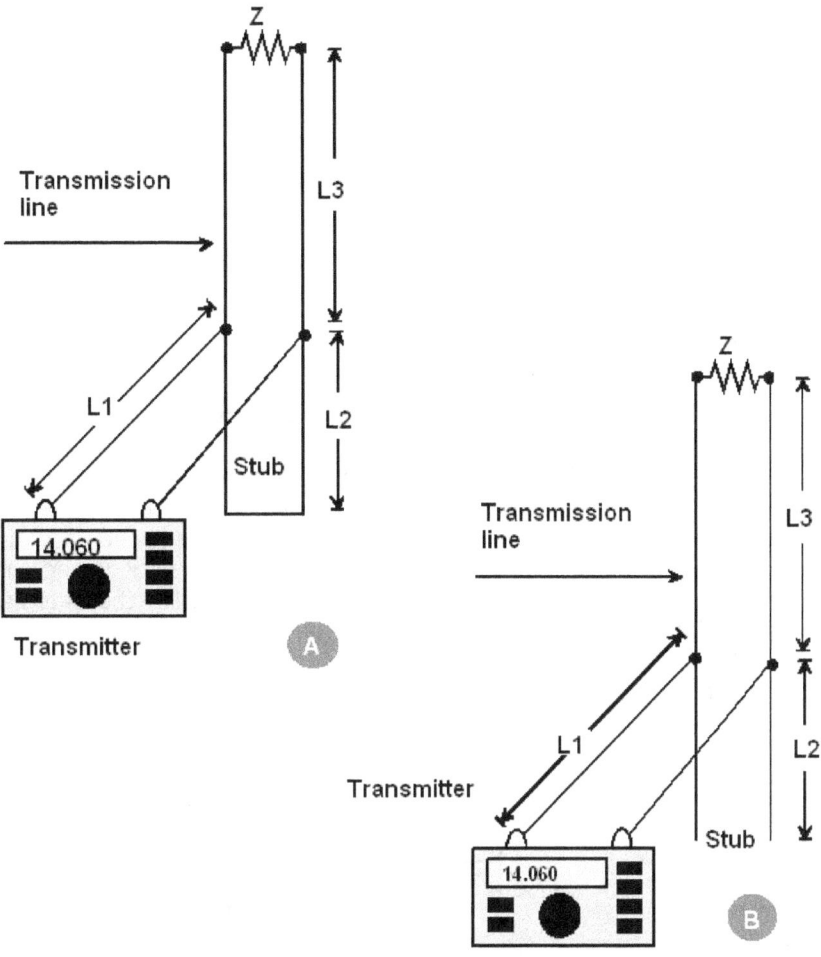

Figure 4 Method of Matching Antenna with help of the Length of the Transmission Line with Stub

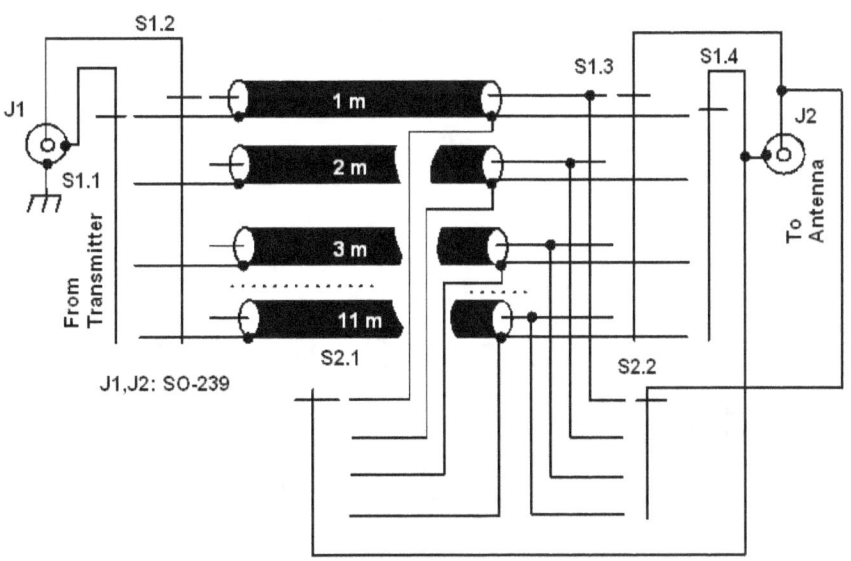

S1: Four Position Eleven Selected Rotary Switch

S2: Two Position Eleven Selected Rotary Switch.
The switch has zero selected position,
hen no one length does not connected to the antenna jack.

Figure 5 Modified Simple HF ATU on lengths of Coaxial Cable

Converting Antenna Tuner MFJ-962D for Operation with Symmetrical Ladder Line

Viktor Drobot, RK3DL

For operation in the Air at all HF- Bands I use to antenna Delta. The antenna is fed by 300- Ohm Ladder Line. To match the antenna with my transceiver I use to ATU MFJ-962D. The ATU has symmetrical transformer at output. The transformer could provide good symmetrical operation … but with antennas that has low reactance. My Delta has significant reactance through amateur's bands. So the concept is not for me.

However it is very simple convert the tuner for operation with symmetrical antenna that has reactance. **Figure 1 A** shows simplified schematic of the ATU MFJ-962D. **Figure 1 B** shows converting ATU for operation with 300-Ohm ladder line that is fed antenna with significant reactance.

73! RK3DL

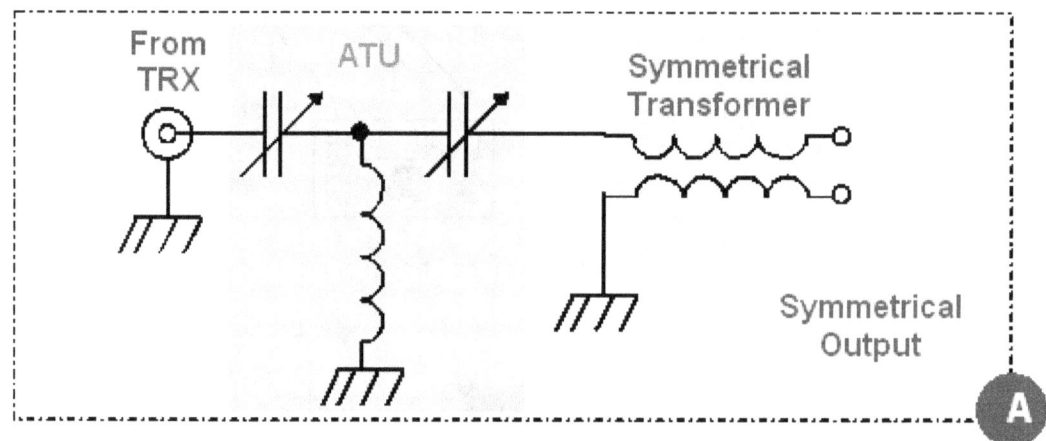

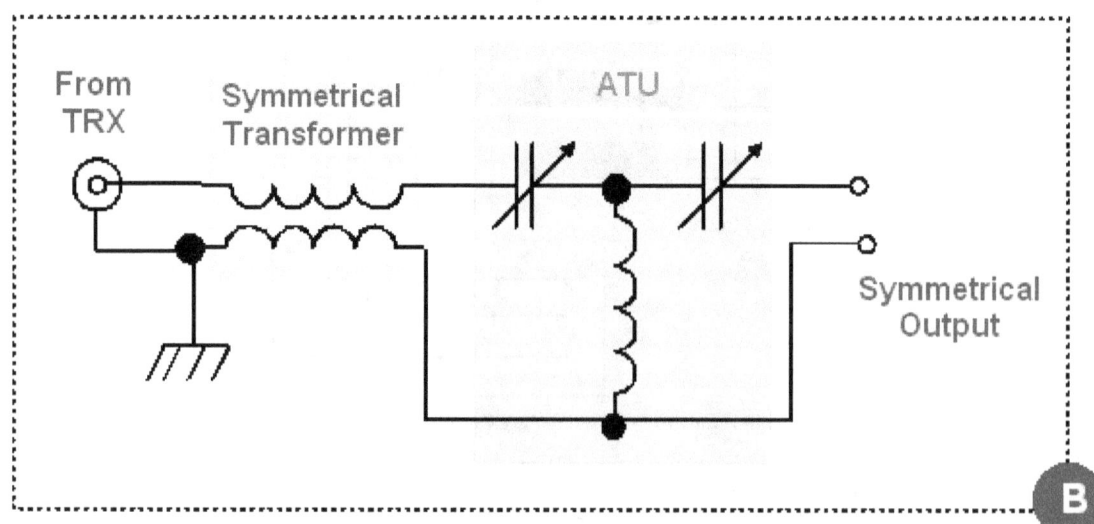

Figure 1

Pocket Antenna Tuner

Boris Popov, UN7CI

This article is described a small (almost pocket) Antenna Tuner that can work with 100- Watt transceiver.

The antenna tuner is a small version of the legendary "Ultimate Transmatch" introduced by Lew McCoy, W1ICP. However at the Lew McCoy's transmatch is used a roller inductor and all capacitors are variable ones. It is very nice for matching but it is not real for pocket design. At this version the roller inductor changed to tapped one and a coupling variable capacitor changed to row of the fixed ones. **Figure 1** shows schematic of the tuner.

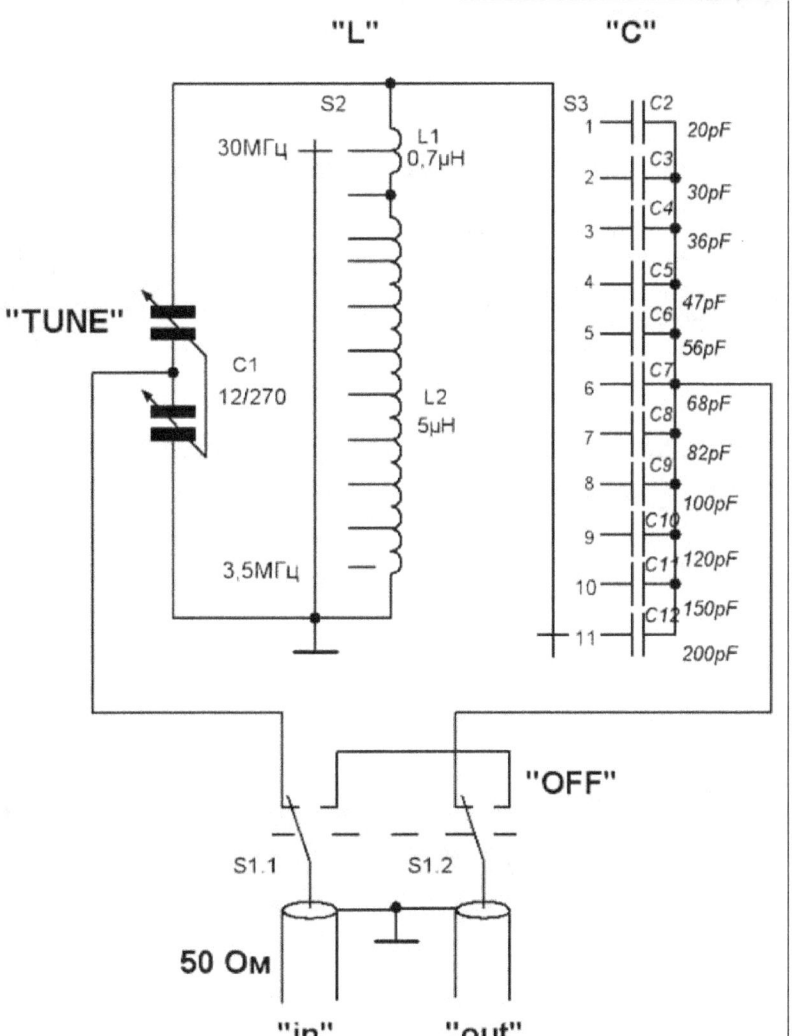

Figure 1 Pocket Antenna Tuner

Antenna and transmitter sockets (to decrease the sizes of the box) did not use at the design. SWR- meter (or RF-Meter) as well did not used here for the reason. Variable tank capacitor had simple inbuilt vernier 1:3. The tuner could match antenna impedance up to 300-Ohm. Limitation is only to working voltage of the row capacitors and input variable capacitor. **Figure 2** shows (for reference) schematic of the Lew McCoy's "Ultimate Transmatch."

Note from I.G.: The Ultimate Transmatch was described in the "Beginner and Novice" section of the July 1970 QST (Page 24). The circuit was very popular that it was also published in several of the ARRL Handbook from the 1970s. I have seen one "Real McCoy Transmatch" (as a seller sad to me) at one of Ontario Hamfests. When I decided if I heed to buy this one or not, another person bought it.

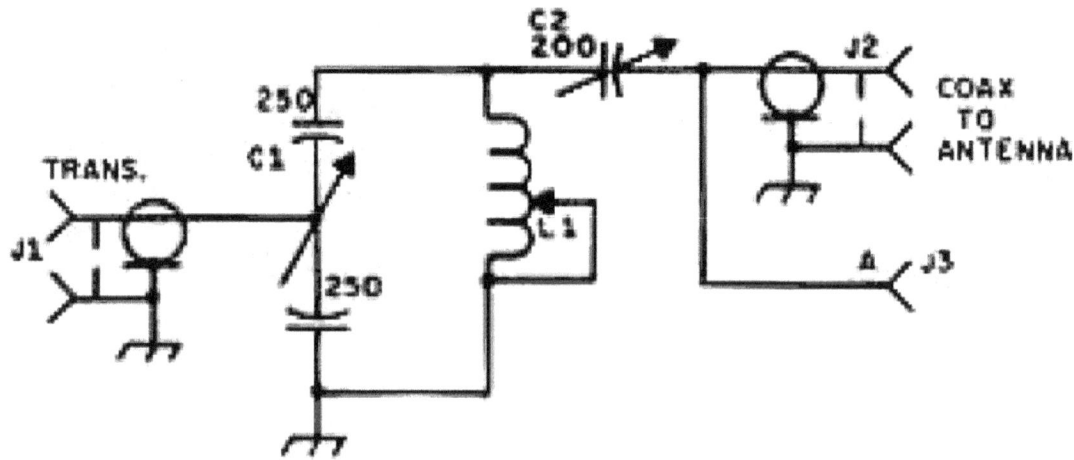

Figure 2 "Ultimate Transmatch" by Lew McCoy, W1ICP

Simple broadband transformer, connected to the tuner, allows use the tuner with symmetrical antennas fed through ladder line and with generation of the wire antennas. The transformer provides 1:4 and 1:9 transformation ratio.

Lew McCoy, W1ICP

Figure 3 Broadband Transformer

Figure 3 shows schematic of the transformer. Transformer is wound on to ferrite ring in 30- mm OD and permeability 20 by triple wires in diameter 1- mm (18-AWG) in Teflon insulation. Pictures show design of the tuner.

Parts List

C1: Variable Capacitor 12/495- pF from old tube receiver.
C2- C12: Ceramic Capacitors, 250- V
S1: Toggle Switch.
S2, S3: Small Rotary Switch for 11- position.
L1: Coiled on to plastic ring (from plastic water-pipe tube) in diameter 20- mm and height 8- mm. Contains 15 turns, tap from the middle. It was used wire in diameter 1.5- mm (15- AWG).
L2: Coiled on to plastic ring (from plastic water-pipe tube) in diameter 20- mm and height 40- mm. Contains 32 turns, tap made from each forth turn. It was used wire in diameter 0.8- mm (20- AWG).

Antenna matching very conveniently may be made in receiving mode. At first, Switch S3 installed at Position 1 (Coupling Capacitor C2). Then with help of C1 and S2 tune tank resonator to resonance (on maxima receiving signals). After that find optimal antenna coupling by S3 and consistently tuning C1 and may be switching S2. May be in transmitting mode the tuner would be need some small tuning on minimal SWR.

73! UN7CI

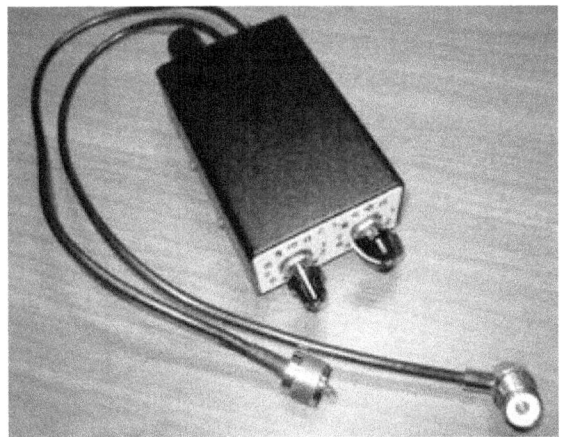

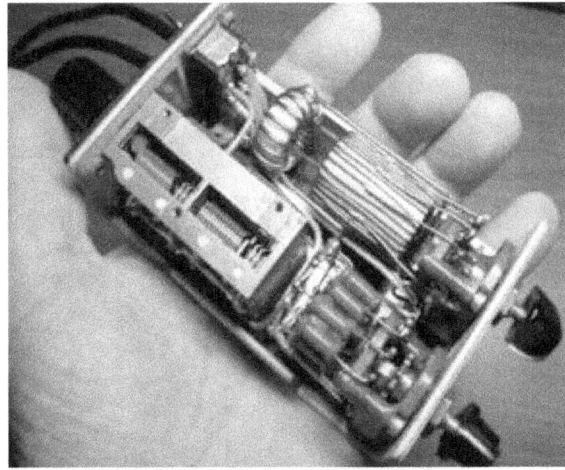

Credit Line for the Article:
www.cqham.ru

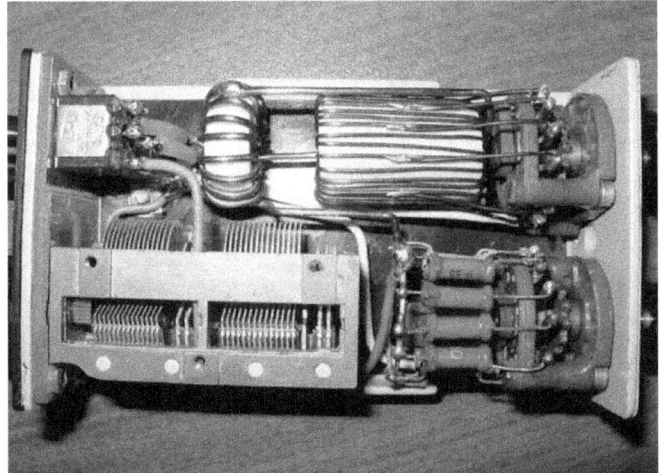

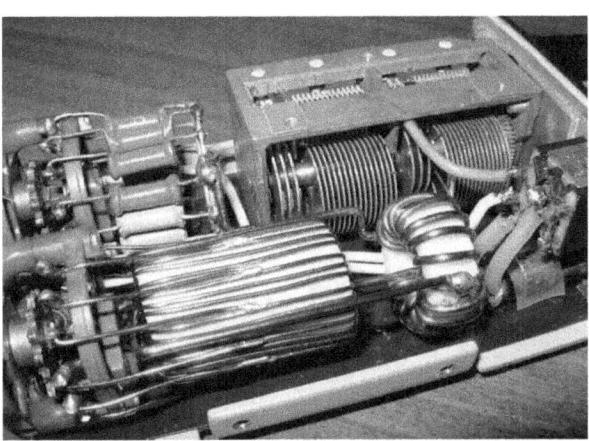

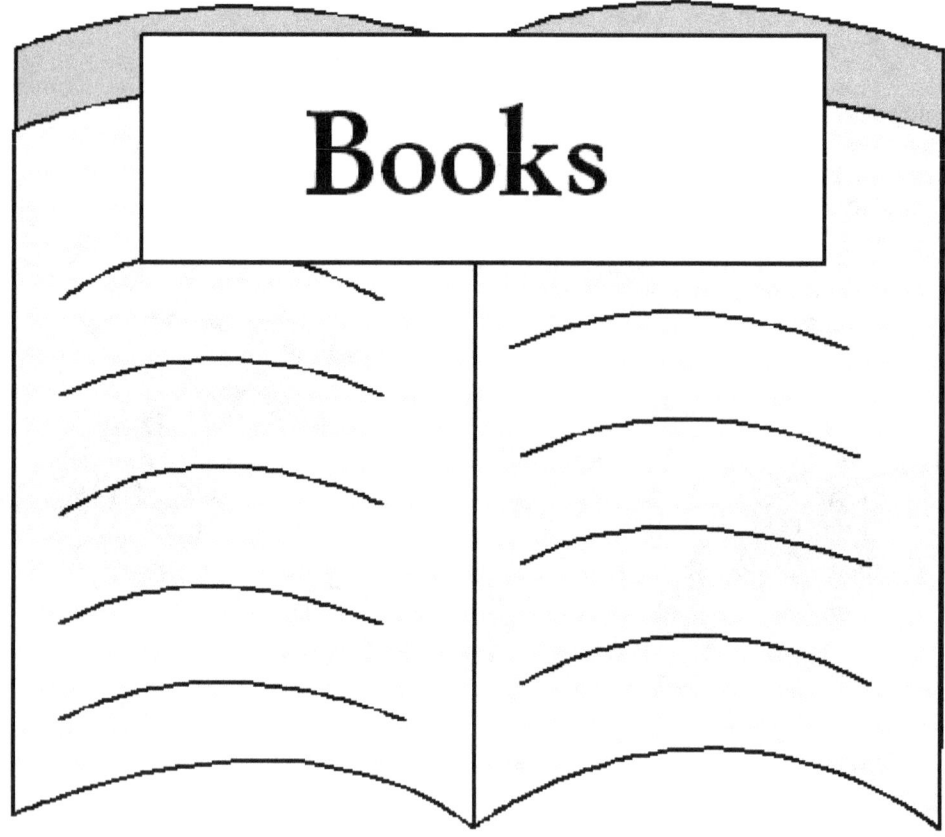

Grounding, Bonding, and Shielding

Excellent Reference to Grounding, Bounding and Shielding. The two books are covered lots practical questions. It is really good books that you can find in the Internet.

From the book: *This document provides basic and application information on grounding, bonding, and shielding practices recommended for electronic equipment.*

It will provide valuable information and guidance to personnel concerned with the preparation of specifications and the procurement of electrical and electronic equipment for the Defense Communications System. The handbook is not intended to be referenced in purchase specifications except for informational purposes, nor shall it supersede any specification requirements.

MILITARY HANDBOOK
GROUNDING, BONDING, AND SHIELDING
FOR
ELECTRONIC EQUIPMENTS AND FACILITIES
VOLUME I OF 2 VOLUMES
BASIC THEORY
404 pages

DISTRIBUTION STATEMENT A. Approved for public release; distribution is unlimited

Link for download:

http://www.antentop.org/library/shelf_grounding_1.htm

MILITARY HANDBOOK
GROUNDING, BONDING, AND SHIELDING
FOR
ELECTRONIC EQUIPMENTS AND FACILITIES
VOLUME II OF 2 VOLUMES
BASIC THEORY
394 pages

DISTRIBUTION STATEMENT A. Approved for public release; distribution is unlimited

Link for download:

http://www.antentop.org/library/shelf_grounding_2.htm

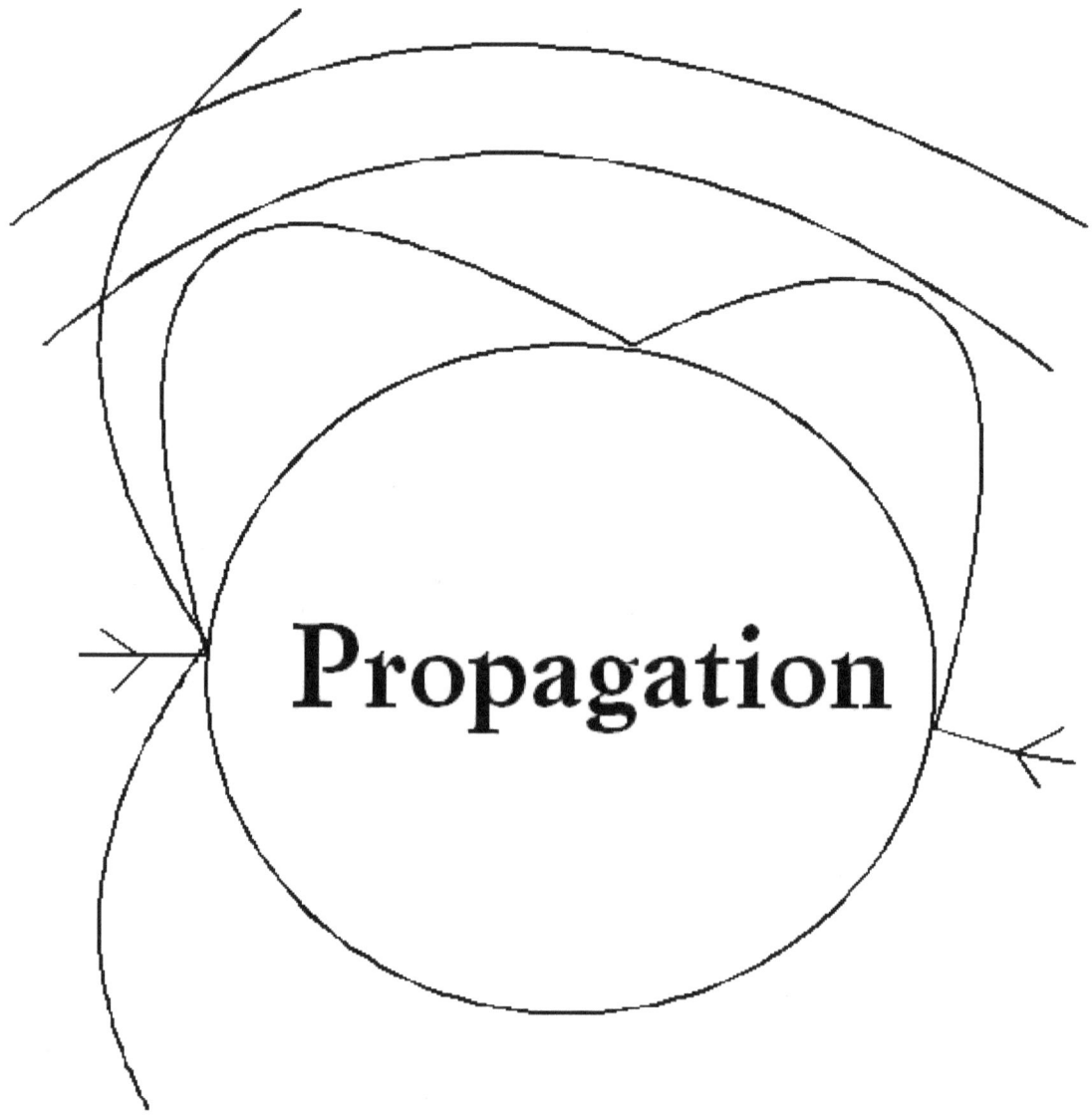

Bridge Effect

By: Igor Grigorov, va3znw

Every day when I drive my car to and from my job I drive under bridges and arches. **Figure 1** shows a bridge. **Figure 2** shows an arch. Inside my car I usually listen to radio. My favorite radio is 680 News Radio. The radio station works on 680- kHz. This radio transmits useful news for me. It's the weather, what is and what will be news in Toronto and the World, as well as local traffic, which road is open, what the road is closed due to an accident construction. Knowledge of the traffic saves me a lot of time.

Figure 1 Bridge

Figure 2 Arch

I noticed that volume of the 680-News usually is changed when I drove under a bridge or under an arch.

ARCH

As usual the level of volume is drop down when I drive under an arch. There are some arches (very small quantity) that do not affect to receiving. **Figure 3** shows area of decreasing reception under the arch. Arch does decreasing of reception for any station at MW Band. Arch does not affect to station at FM-Band.

The arch has stable behavior in influence to the MW station. If an arch does decreasing in reception so the arch always does the decreasing. The effect exists in summer, in winter, in rain, in snow, in traffic and on empty road.

If an arch does not influenced to radio reception the arch always does not influenced to this one. It is right for summer, for winter, for rain, for snow, for traffic and for empty road.

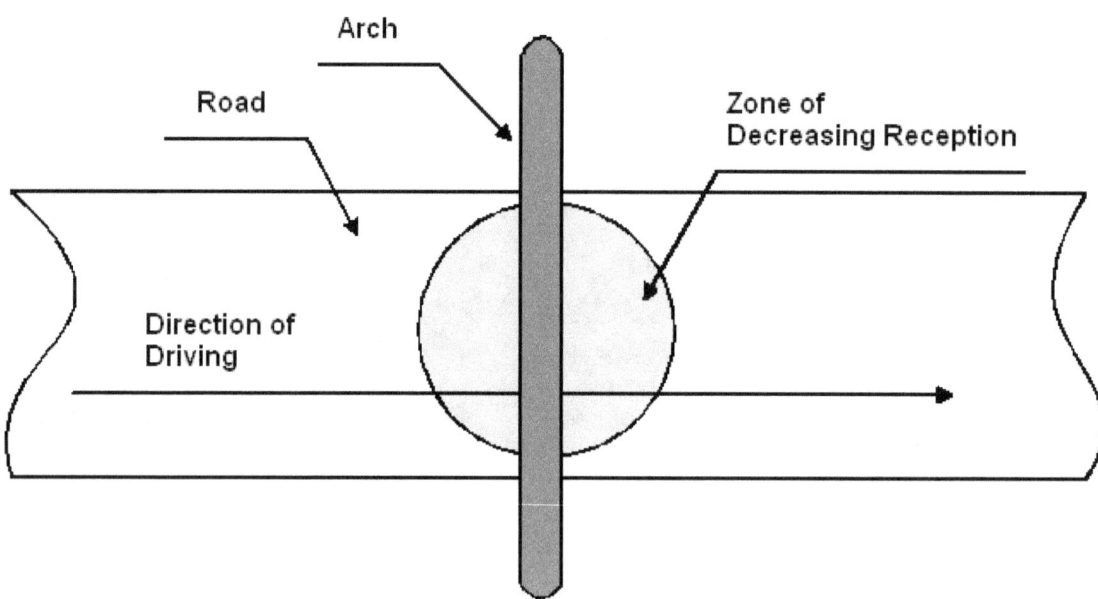

Figure 3 Area of decreasing reception under the Arch

Generally speaking it is possible to find explanation for the "arch effect."

It could be assumed that the arch is a closed loop, which in its zone of influence produces local shielding, or shift in direction of propagation of radio waves. Both of this effect could result to decreasing in receiving of the MW radio. Automatic Gain Control cannot stand the fast decreasing of the signal level.

The arch, which does not affect the reception of MW signals, is not a closed loop which can screen or modify locally direction of propagation of radio waves in MW Band.

BRIDGE

The bridge influences to radio reception in MW Band in more complicated way compare to the arch. It is possible describe three typical cases and four untypical ones. Bridge does not affect to station at FM-Band.

Three typical cases of the bridge behavior

First typical case is shown in **Figure 4**. It is the most typical case in influence of the bridge to radio in MW Band. Level of volume is drop down when car is driving under the bridge and after the car driving up to 20- 50- meters out of the bridge. Then the volume goes back to usual level.

Second typical case is shown in **Figure 5**. Level of volume is drop down when car is driving strictly under the bridge. When car is driven out of the bridge the volume goes back to usual level.

Third typical case (but rare compare to the first two cases) is shown in **Figure 6**. Level of volume is drop down when car exits out of the bridge.

That is interesting that picture of attenuation of the radio reception lay in the direction of the driving. If a car would drive under the same bridge in the opposite direction the picture (**Figure 4, 5, 6**) would be the same. In my opinion the picture of attenuation of MW reception may be explained by various degrees of shielding of the MW radio by bridge and inertia in the receiver's AGC. .

Behavior of bridges (to influence to MW radio) is unstable. It is oddly enough but any bridge may behave differently. One day, its behavior is consistent with **Figure 4**, in another day its behavior is consistent with **Figure 5**, and the next day its behavior is consistent with **Figure 6**.

Four untypical cases of the bridge behavior

First untypical case of the bridge behavior is that the bridge has no effect on the reception in MW Band.
For the *Second*, *Third* and *Fourth* untypical case I take inversion in the influence to the MW reception. It means that instead of decreasing in level of volume of MW reception there are increasing in level of volume of MW reception.

Picture of the increasing is the same as for picture of the decreasing of MW reception. So it is the same as shown on **Figures 4**, **5**, **6**. Only the volume is increased instead decreasing as shown at the **Figures**.

What it depends on?

I tried to find factors that cause the typical and untypical cases in bridge behavior. I took in consideration seasons, weather, traffic, Moon cycles- but I could not.

At the same time of year, at the practically the same weather and traffic, today bridge did decreasing in volume of the MW radio, next day the bridge did increasing in the volume, and the next day did no influence to the reception.

I noticed that the same effect is usually observed for 3- 5 bridges in one direction. So, if one bridge did decreasing in reception so as the rule the next 3- 5 bridges also did decreasing in reception of MW radio.

If one bridge did increasing in reception so as the rule the next 3- 5 bridges also did increasing in reception of MW radio.

There are some bridges that have stable effect. The bridges always did decreasing in reception of MW radio or did not influence to the reception.

However I newer did not find bridges that stable increased the radio reception in MW Band.

These effects I have observed at all cars that my family had. There were Nissan Sentra, Saab, Chevrolet Malibu. This effect I observed at our recent cars- Chevrolet Aveo and Sonic.

73! va3znw

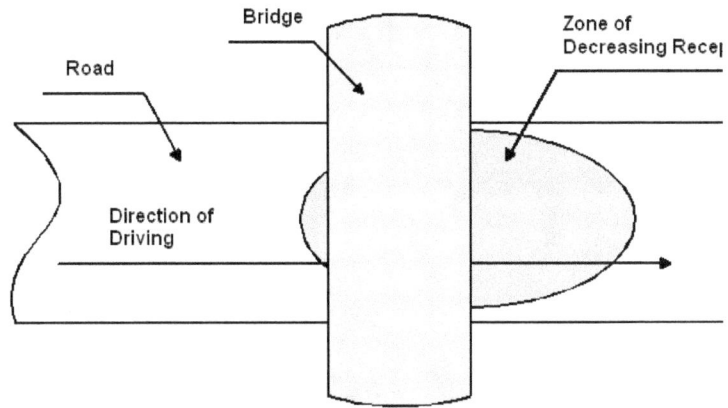

Figure 4 Typical zone of attenuation under the bridge

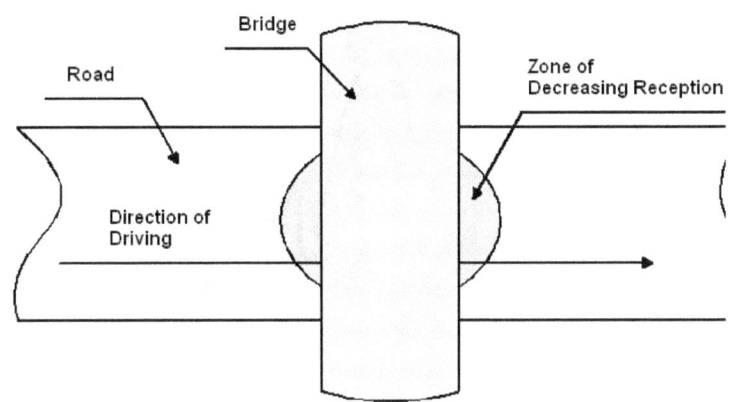

Figure 5 Zone of attenuation strictly under the bridge

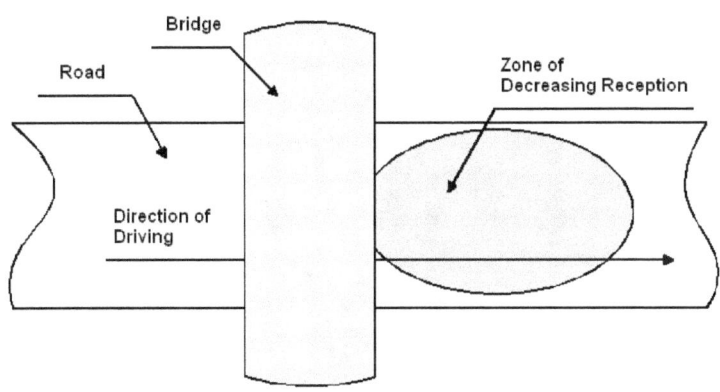

Figure 6 Zone of attenuation at the exit out of the bridge

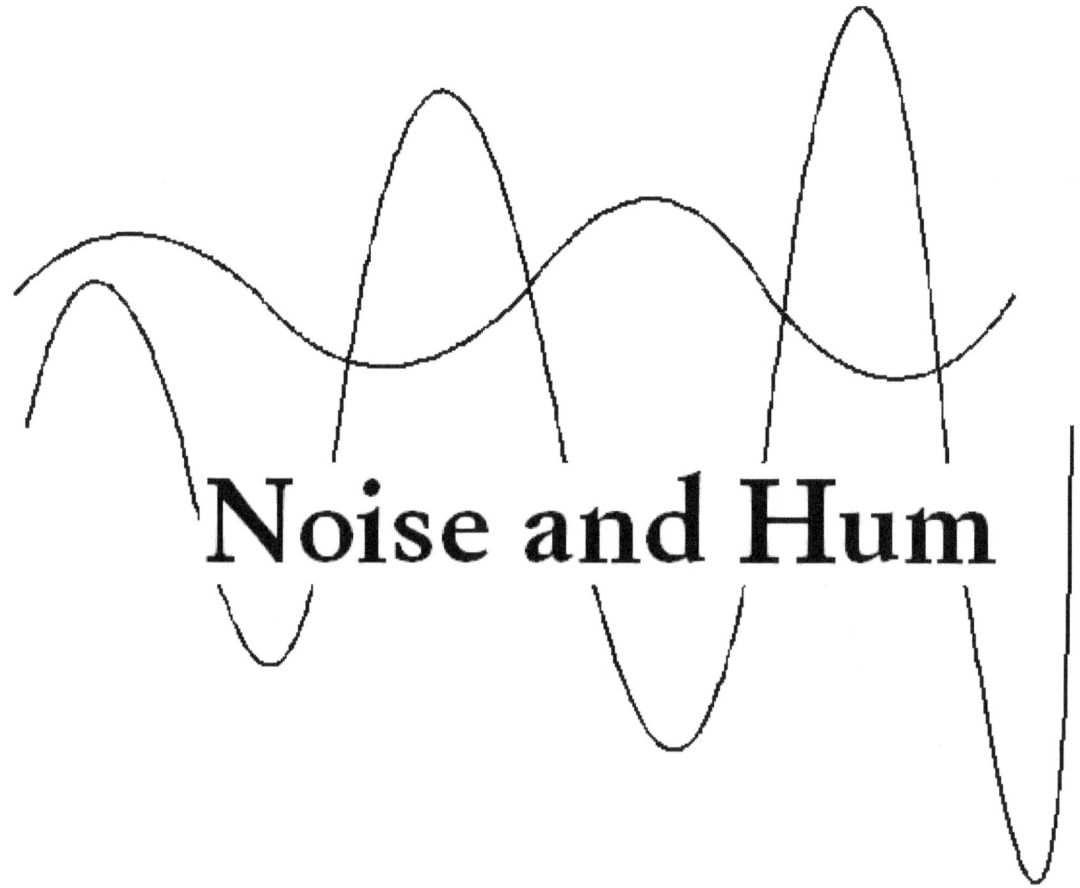

Nature of Hum

By: Michael J Hebert, NH7SR

Credit Line: Forum QRP- L

Some 10 years back I was doing a lot of spherics recording using both E-field probes and loops. The VLF spectrum is, of course, absolutely saturated with hum from the power grid. There were some interesting things I learned about the nature of electrical grid hum during that time. More or less as follows in no particular order....

1) The primary frequency (60Hz in the US) and the third harmonic have roughly the same "signal" strength since we use 3-phase power distribution. 2nd harmonic levels are considerable lower in level but sound much dirtier.

The next highest level that I encountered in my recordings was at 720Hz using either an E-field probe or a loop sensor. The fundamental sounds rough, the third harmonic sounds fairly clean and 720Hz sounds clean.

2) The E-field probe was more sensitive to the higher harmonics than the loop sensor... probably mostly due to circuitry differences in the amplifiers. The E-field probe became less sensitive to the fundamental and lower harmonics they higher it was held.

3) Using the E-field probe I found the noise envelope surrounding tall buildings tended to have a radius approximating the height of the building. With a loop sensor I could be much closer to a building before the hum level equalled that picked up the E-field probe.

I attribute this to the fact that E-field strength decreases as the square of the distance from the source whereas the magnetic field decreases as the cube of the distance.

4) With the loop sensor I could rotate the loop to partially null hum pickup but only if the loop was mounted or held upright. If, OTOH, it was held or mounted with the turns parallel to the earth it was easily saturated by ground-conducted hum currents even when located several hundred feet from tall buildings.

In certain areas I could trace out the apparent paths being traversed by the earth currents.

73/72! **NH7SR**

Note from VA3ZNW:

Just my 2 pens added. In the 80- 90-s I actively explored simple DC and regenerative receivers. Main luck of the receivers was the hum. The receivers pick up the hum (50-Hz in Russia) ever without an antenna. Ever the receiver was fed from a battery. Any one receiver- transistor or tube one.

It was impossible to eliminate the hum. I used rejected filters at input and between stages at Audio Amplifier. I used Audio Amplifiers that did not work below 300- Hz. No success! Receiver roared at the higher harmonics. It was in my city shack.

But when I try the receivers at a field conditions, far away from the city and any electrical distribution wires I was amusing.

The first my impression was- the receiver did not work. No any hum. But then when turn around tuning capacitor I heard stations. Lots stations that I cannot receive at the city. I understand, that the receiver works fine, just no any hum. No lots of intermodulation interferences.

It was very amusing! Receiver that at the city conditions roared and noised the same receiver in the field gave very good reception without any noise. However, as I noticed, at cloudy and rainy weather I could pick up the electrical hum. However the hum was not strong as it was in the city. As well it was appeared intermodulation interferences.

So, DC and regenerative receiver do not like city with the hum from main from electrical power equipment and modern electronic devices. My nostalgia is the times when I received interferences only from power tube horizontal generator of the TV.

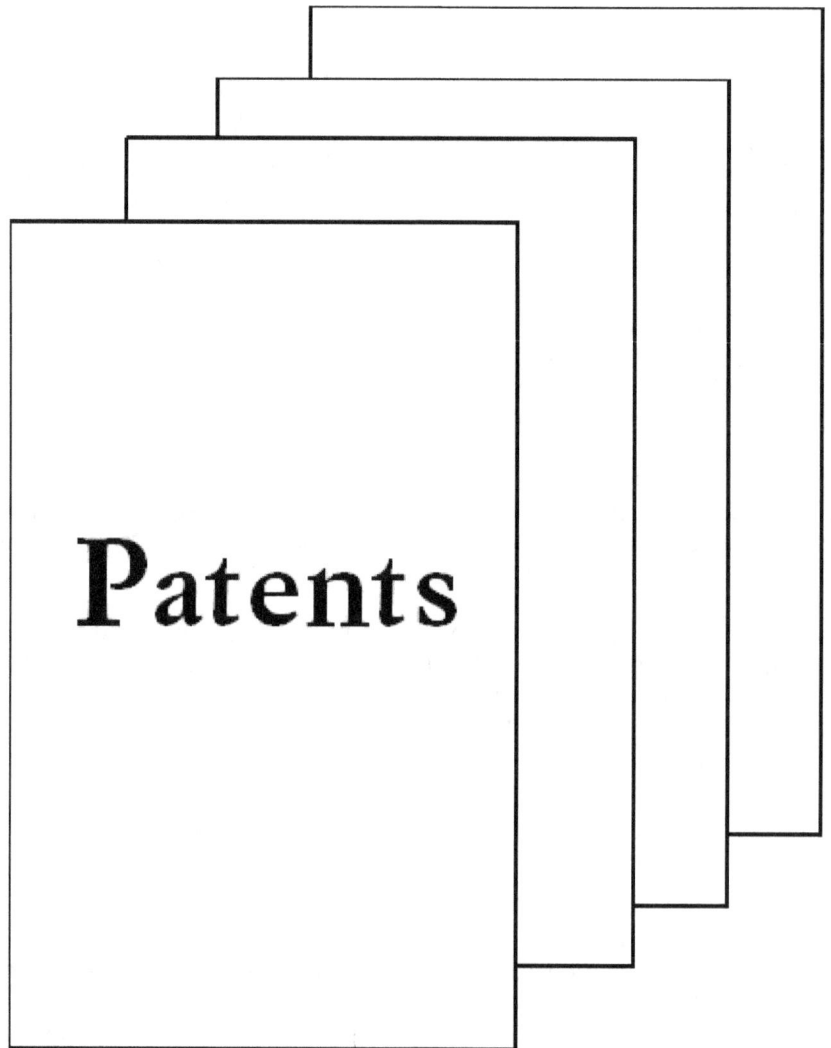

www.antentop.org

Antenna for Mobile Communications

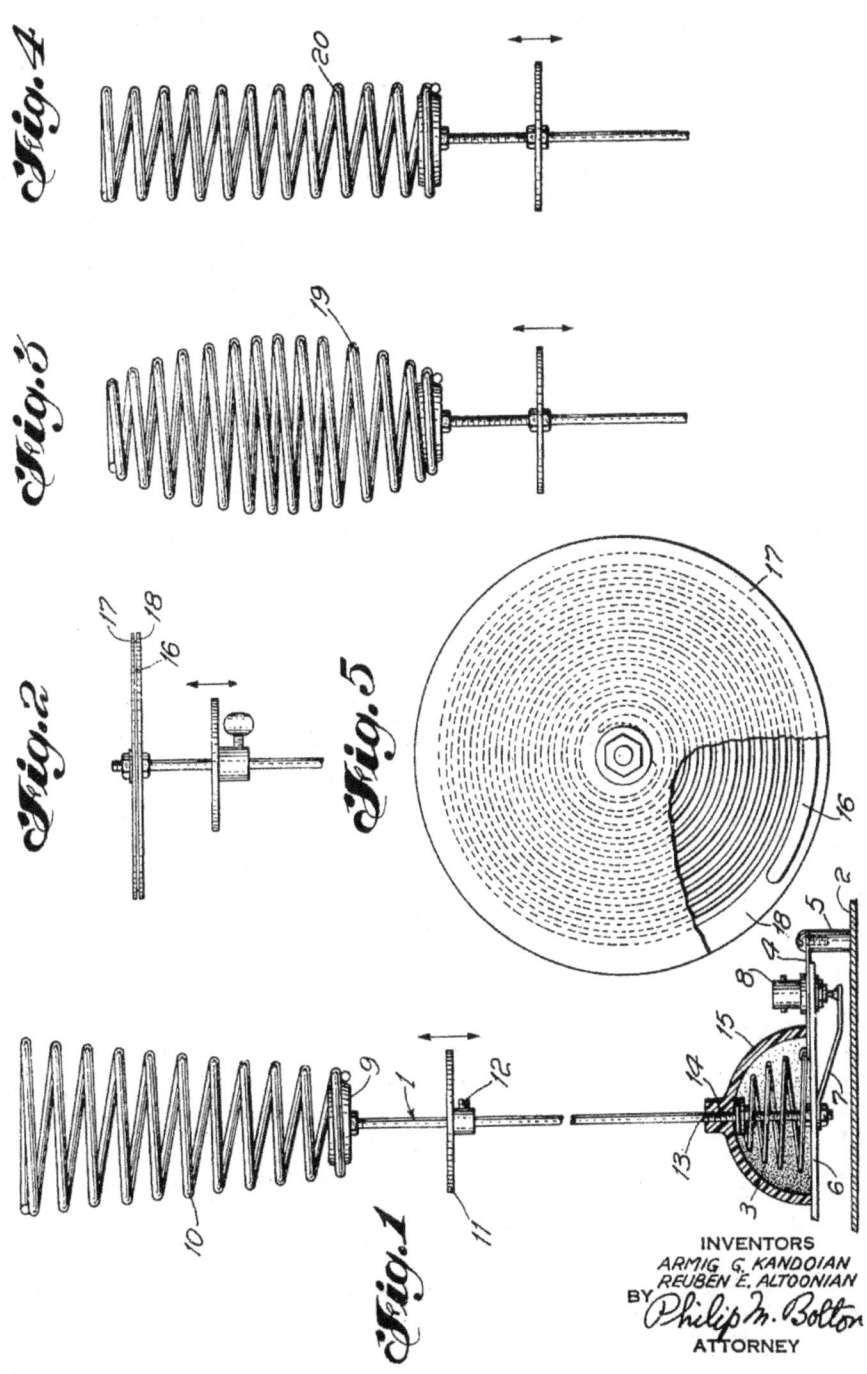

United States Patent Office

2,850,732
Patented Sept. 2, 1958

2,850,732

ANTENNA FOR MOBILE COMMUNICATIONS

Armig G. Kandoian, Glen Rock, and Reuben E. Altoonian, Millington, N. J., assignors to International Telephone and Telegraph Corporation, Nutley, N. J., a corporation of Maryland

Application October 3, 1955, Serial No. 538,038

4 Claims. (Cl. 343—752)

This invention relates to antennas for use on mobile equipment and particularly vertical grounded antennas using top loading.

Vertical antennas on mobile equipment, such as for example automobiles, tend to be so long around frequencies of 20 to 50 megacycles that they are difficult to mount. For example, in the 30 to 50 mc. band, vertical grounded antennas of 70 to 90 inches are standard. It has been the practice to mount such antennas on the fender or rear deck of the automobile and off-center with respect to the body metal which acts as the ground plane. The result of this has been to adversely affect the desired omni-directional characteristics of such antennas.

It has heretofore been suggested that the physical length of vertical antennas could be decreased by using "top loading." This consists essentially of adding lumped circuit elements to the top of the antenna. In the prior art, these elements have consisted of the combination of a lumped-constant coil inductance plus the capacitance of a separate body, which had the form of a sphere or disk. Antennas of these types are described in The A. R. R. L. Antenna Book, 1949 edition, published by The American Radio Relay League, page 61, and in the copending U. S. application of A. G. Kandoian for "Antenna," filed June 21, 1954, bearing Serial No. 438,222.

Such top-loaded antennas have not been adopted to any appreciable extent and this in part may be attributed to the complexity of such antennas and the mechanical difficulties in constructing them. Such antennas must be quite rugged and designed to present a minimum of wind resistance. At the same time, they must be relatively inexpensive. It is also important to provide some means for tuning such antennas and some means to match the antenna impedance to its load or source.

An object of the present invention is the provision of an improved vertical grounded antenna of the top-loaded type for use in mobile equipment. Some of the features of this antenna include its simplicity, high Q, and compactness. The length of one example of the present invention, operating in the 30 to 50 mc. band, is 16 to 18 inches, as opposed to 70 to 90 inches for the standard vertical ground plane antenna.

Another object of the present invention is the provision of an antenna of the type hereinbefore described having simple means for tuning said antenna.

Another object of the present invention is the provision of an antenna of the type hereinbefore described which presents small wind resistance and is rugged in structure.

In carrying out one aspect of the present invention, the top loading of the antenna is produced solely by a single spiral coil (which may be a flat or helical spiral) which provides both inductance and a substantial capacitance-to-ground. In accordance with a further feature of the present invention, the antenna is tuned by changing the position of a disk mounted on the vertical portion of the antenna and below the spiral so as to vary the loading effect of the spiral.

Other and further objects of the present invention will become apparent, and the foregoing will be better understood with reference to the following description of embodiments thereof, reference being had to the drawings, in which:

Fig. 1 is a schematic diagram of a vertical grounded antenna according to the present invention;

Figs. 2 through 4 are similar views of modified portions of the antenna arrangement of Fig. 1; and

Fig. 5 is a schematic top plan view of the top portion of the antenna illustrated in Fig. 2.

Referring now to the embodiment illustrated in Fig. 1, the tuned dipole antenna system or vertical antenna illustrated there comprises a vertically disposed conductive tube or rod **1** which serves as the radiator. The vertical radiator **1** is terminated to ground **2** through an impedance matching inductance **3**, connector strap **4** and conductive supporting post **5**. The bottom of rod **1** is fixed in an insulating plate **6** and connected via a lead **7** to the inner conductor of a coaxial cable connector **8**, the connector in turn connecting the antenna to a transmitter or receiver (not shown). The ground plane or base **2** may be the top of an automobile or other mobile equipment with the antenna mounted at the geometrical center thereof. The length of the vertical radiator **1** from its top, to the top of coil **3** is considerably less than a quarter wavelength and to effectively render this antenna a quarter wavelength, "top loading" is employed. For this purpose the top of the vertical radiator **1** has a mounting disk **9** on which a tapered helical spiral **10** is supported whose outer configuration is in the form of a truncated inverted cone, the central vertical axis of the cone being aligned with vertical radiator **1**. The spiral is made of heavy gauge wire which may be as thick as the vertical radiator **1**, or preferably at least half the diameter of the vertical radiator **1**, so as to provide a rugged structure. Moreover, the spiral **10** is made of relatively large dimensions so that not only does the spiral provide inductance, but it also provides a relatively large capacitance-to-ground. In this embodiment, unlike that of the prior art, no separate means is required to provide the desired capacitance.

In accordance with another feature of the present invention, the antenna described in Fig. 1 is tuned by means of a conductive plate or disk **11** surrounding the vertical radiator **1** and extending transversely thereto. The disk **11** is adjustable along the length of the radiator **1** and is held in position by simple means, such as a set screw **12**. Varying the position of disk **11** varies the resonant frequency. This provides an extremely simple way of adjusting the resonance of the antenna to the mid-frequency of the band over which it is to operate.

In the example referred to hereinabove, a disk having a diameter of 2″ was employed. It is to be noted that this is considerably less than the maximum diameter of the spiral measured in its width. The thickness of the disk need only be sufficient for mechanical strength.

The vertical radiator **1** may be additionally supported in any suitable way. For example, the vertical conductor **1** may be provided with a threaded portion **13** feeding into the complementary thread **14** of a dielectric member **15** in the form of a hemisphere, which member **15** is mechanically supported above the ground plane by suitable means, such as legs and screws or the like (not shown). Other forms of support for the antenna will be obvious.

2,850,732

3

Typical dimensions for one embodiment of the antenna of Fig. 1 for use at 30 to 50 mc. are as follows:

	Inches
Height of spiral	4
Maximum diameter (width) of spiral	3
Height of vertical rod 1 from top thereof to top of hemisphere 13	10½
Distance from point 13 to base	1½

Referring now to the embodiment illustrated in Figs. 2 and 5, this structure differs from that described in Fig. 1 in that the top-loading spiral is a flat spiral 16, as can be seen clearly in Fig. 5, and is supported between two dielectric sheets 17 and 18. The spiral 16 and sheets 17 and 18 extend transversely to the vertical radiator 1. The overall dimensions of this antenna may be approximately the same as those of Fig. 1 although the diameter of the spiral may be a little larger, for example 4″. Tuning of the antenna and the other details thereof outside the spiral configuration are the same as in Fig. 1. It is to be noted that no additional capacitance is required with spiral 14 providing all the desired capacitance-to-ground. The spiral may be made of flat wire flattened at the top and bottom with the width of the wire extending in the same plane in which the spiral lies. The width of the spiral wire is preferably at least half that of the vertical radiator so as to provide a substantial capacitance-to-ground. The outermost turn of the spiral 14 may be made wider than that of the other turns to provide additional top capacity.

In the embodiment illustrated in Fig. 3, a helical spiral 19 is employed whose outer configuration is in the form of a prolate spheroid whose capacitance-to-ground and inductance serve to load the antenna. On the other hand, in Fig. 4, the spiral 20 which serves to load the antenna is cylindrical in outer configuration. The thickness of the wire of these spirals is preferably the same as that of the spiral of Fig. 1 and the diameter of each of these spirals is preferably at least 3″. In all the embodiments shown, no additional capacitance-to-ground element is necessary. The transverse disk tuning means is employed in each and the base coil for matching the impedance of the antennas is provided in each antenna unit.

4

While we have described above the principles of our invention in connection with specific apparatus, it is to be clearly understood that this description is made only by way of example and not as a limitation to the scope of our invention as set forth in the objects thereof and in the accompanying claims.

We claim:

1. An antenna system comprising a vertically disposed radiator, means for coupling a transmission line to said radiator, impedance matching means coupled to said vertical radiator, means positioned near the top of said vertical radiator for increasing the effective electrical length of said vertical radiator, and means for tuning said antenna system comprising a planar conductor mounted beneath said antenna loading means and adapted to be moved vertically.

2. An antenna system acording to claim 1, wherein said planar conductor comprises a disk mounted on said vertical radiator and extending transversely thereof.

3. An antenna system comprising a vertically disposed radiator, means for coupling a transmission line to said radiator, impedance matching means coupled to said radiator, means for increasing the effective electrical length of said vertical radiator comprising solely a spiral coil disposed toward the top of said vertical radiator with one end attached to the vertical radiator, said coil having a substantial inductance and being dimensioned to have a substantial capacitance-to-ground, and means for varying the tuning of the antenna system comprising a planar conductor slideably mounted on said vertical radiator for vertical movement towards or away from said spiral coil.

4. An antenna system according to claim 3 in which said planar conductor is in the form of a disk through which said vertical radiator passes, said disk extending transversely to said radiator.

References Cited in the file of this patent

UNITED STATES PATENTS

2,319,760   Becwar _____ May 18, 1943

OTHER REFERENCES

Air Force Manual 52–19, Antenna Systems, pages 119–121 incl.

Armig G. Kandoian (S'35–A'36–SM'44) was born in Van, Armenia, on November 28, 1911. He received the B.S. degree in 1934 and the M.S. degree in electrical communication engineering in 1935, both from the Harvard University. Since 1935, Mr. Kandoian has been with the International Telephone and Telegraph Corporation and associated companies. His work has been primarily developments dealing with antennas, radiation, measurements, link communication, and air navigation. He is at present head of the radio and radar components division of Federal Telecommunication Laboratories.

Mr. Kandoian received the honorable mention award in the Eta Kappa Nu recognition of outstanding young electrical engineers for 1943. He is a member of Tau Beta Pi, Harvard Engineering Society, and the American Institute of Electrical Engineers.

ANTENTOP

ANTENTOP is *FREE e- magazine*, made in **PDF**, devoted to antennas and amateur radio. Everyone may share his experience with others hams on the pages. Your opinions and articles are published without any changes, as I know, every your word has the mean.

A little note, I am not native English, so, of course, there are some sentence and grammatical mistakes there… Please, be indulgent!

Publishing: If you have something for share with your friends, and if you want to do it *FREE*, just send me an email. Also, if you want to offer for publishing any stuff from your website, you are welcome!

Copyright: Here, at ANTENTOP, we just follow traditions of *FREE* flow of information in our great radio hobby around the world. A whole issue of ANTENTOP may be photocopied, printed, pasted onto websites. We don't want to control this process. It comes from all of us, and thus it belongs to all of us. This doesn't mean that there are no copyrights. There is! Any work is copyrighted by the author. All rights to a particular work are reserved by the author.

Copyright Note: Dear friends, please, note, I respect Copyright. Always, when I want to use some stuff for ANTENTOP, I ask owners about it. But… sometimes my efforts have no success. I have some very interesting stuff from closed websites however their owners keep silence… as well as I have no response on some my emails from some owners

I have a big collection of pictures. I have got the pictures and stuff in different ways, from *FREE websites*, from commercial CDs, intended for *FREE using*, and so on… I use to the pictures (and seldom, some stuff from free and closed websites) in ANTENTOP. *If the owners of the Copyright stuff have concern*, please, contact with me, I immediately remove any Copyright stuff, or, if it is necessary, all needed references will be made there.

Business Advertising: ANTENTOP is not a commercial magazine. Authors and I (Igor Grigorov, the editor of the magazine) do not get any profit from any issue. But of course, I do not mention from commercial ads in ANTENTOP. It allows me to do the magazine in most great way, allows me to pay some money for authors to compensate their hard work.

So, if you want paste a commercial advertisement in ANTENTOP, please contact me.

Book Advertising: I believe that *Book Advertising* is a noncommercial advertisement. So, Book Advertising is *FREE* at ANTENTOP. Contact with me for details.

And, of course, tradition approach to **ANY** stuff of the magazine:

BEWARE:

All the information you find at *AntenTop website* and any hard (printed) copy of the *AnTentop Publications* are only for educational and/or private use! I and/or authors of the *AntenTop e- magazine* are not responsible for everything including disasters/deaths coming from the usage of the data/info given at *AntenTop website/hard (printed) copy of the magazine*.

You use all these information of your own risk.